Dynamic Thermal Analysis of Machines in Running State

Lihui Wang

Dynamic Thermal Analysis of Machines in Running State

 Springer

Lihui Wang
Department of Production Engineering
KTH Royal Institute of Technology
Stockholm
Sweden

ISBN 978-1-4471-5878-3   ISBN 978-1-4471-5273-6   (eBook)
DOI 10.1007/978-1-4471-5273-6
Springer London Heidelberg New York Dordrecht

Printed on acid-free paper

Springer is part of Springer Science+Business Media (www.springer.com)

Preface

Machine tool design and manufacturing are tightly associated with innovation. They have been the key areas that support and influence a nation's economy since the eighteenth century. This is especially true for the machine tool industry. As the mother machines, machine tools form the basis for the development of many other products. In the past centuries, machine tool design and manufacturing have contributed significantly to modern civilisation and created the momentum that drives today's economy. Despite various achievements, we are still facing challenges due to the growing complexity in machine tool design and development.

The complexity comes from multiple directions. On one hand, the complex shape and geometry of new products require more sophisticated machining capability of machine tools to produce the new products to specifications. On the other hand, the ever-growing quality requirements of customers demand tight tolerance for the manufactured products. Consequently, a machine tool must perform accurately, reliably and with high repeatability over time.

It is a common practice that a machine tool is designed and analysed in static state with constraints and under assumptions. A so-designed machine may not perform consistently over time in running state. Among varying working conditions in reality, uneven temperature distribution across the machine components is most influential to machining accuracy due to thermal deformation during metal cutting operations. This thermal deformation alters the cutter–workpiece relationship dynamically, resulting in unsatisfactory machined surfaces. Although modern computer-aided techniques are helpful in addressing some of the problems in machine tool design, there still remains a big gap between the required performance and the real performance of a machine because of the static nature of problem-solving capability of current software tools and applied technology. Two significant problems remain to be solved: (1) how to effectively share information between solid modelling and engineering analysis modules, and (2) how to satisfactorily enable dynamic analysis to simulate the true behaviours of machine tools in running state.

To bridge the gap and present the state-of-the-art of machine tool design and development to a broad readership, from academic researchers to practicing engineers, is the primary motivation behind this book.

This book summarises basic concepts, fundamental considerations and problem-solving algorithms relevant to machine tool design and analysis in running state. The book is composed of six chapters, and a brief outline of each chapter is given below. Chapter 1 provides an introduction of the historical background of relevant research, and clarifies why this research is necessary to be carried out and what the objectives of the research are. Based on the literature analysis, the motivation of this book on dynamic thermal analysis for machine design is carefully laid out. Chapter 2 provides a detailed description of the techniques for representation of machine tool models. Constructive Solid Geometry (CSG) is adopted as the primary representation in this book, among contemporary solid modelling methods, for the ease of modelling of machine tool structures. A low-level data structure is documented for establishing a primitive library with the characteristics of machine tool structures. Chapter 3 presents a method to implement a product modelling system for machine tool design. Based on the design process of machine tools, considerable requirements to be met in the product modelling system are specified, together with the definition of product models. For the purpose of system development, a high-level data structure for both components and an entire machine tool is proposed. Kinematic simulations, as case studies, are carried out by using the product modelling system. The connection between the product modelling system and an integrated CAD/CAE system is emphasised.

Moving from modelling to analysis, Chap. 4 presents a new method for dynamic finite element mesh generation, called Coded Box Cell (CBC) substitution approach in this book. Hexahedrons are chosen as the mesh elements for the convenience of automatic generation and modification of FEM meshes, since machine tools are mostly composed of cuboids and box-type primitives. The extension of the CBC substitution to curved objects is introduced by using mapping and inverse mapping techniques. Full details of the CBC substitution approach are described in this chapter. A new data structure of machine tool model for its utilisation in FEM mesh generation is also given, based on which several practical case studies are carried out. Chapter 5 showcases an application of the CBC substitution approach to finite element analysis. Since thermal error is the major factor that affects the machine tool designs, the thermal analysis is taken into consideration. The table and the base of a machine tool is simplified as the model of the thermal analysis. The Emphasis is given to the interpolation of intermediate results between consecutive analytical steps for the purpose of a continuous calculation, when relative motions take place between the table and the base. The corresponding experiments under the same running conditions are conducted. Based on the analytical and experimental results, discussions and evaluations are documented. Finally, Chap. 6 draws to the conclusions of this book. After summarising each chapter and the research findings, challenges and future research directions are pointed out for interested readers working in the same field.

Finally, the author would like to take this opportunity to express his appreciation to Springer for supporting this book, and would especially like to thank Anthony Doyle, Senior Editor for Engineering, and Christine Velarde, Senior

Editorial Assistant, for their patience, constructive assistance and earnest cooperation, both with the publishing venture in general and with the editorial details. We hope that readers find this book informative and useful.

Stockholm, March 2013   Lihui Wang

Contents

Symbols

α_c	Heat transfer coefficient across S_3
c	Specific heat of material
$[C]$	Heat capacity matrix
$d(m, n)$	Distance between nodes m and n
D	Direction of normal vector on the contacting plane of the primitives F and M
$\tilde{D}$	Displacement of relative motion after Δt
δ^F, δ^M	Mesh sizes (lengths between adjacent nodes) of the FEM meshes of the primitives F and M, respectively
δ^{max}	Maximum mesh size of a meshwork
Δt	Time interval of calculation
e	Element
f	Mapping function
F	Fixed primitive
$\{F\}$	Heat flux vector
g	Inverse mapping function
I	Functional
I_{pr}	An image of pr
ID_{CBC}	Set of identification numbers of CBCs to be used for substitution
$[J]$	The *Jacobian* matrix
k	The number of elements
$[K]$	Heat conductivity matrix
L^F, L^M	Node labels of the primitives F and M, respectively
$\lambda_x, \lambda_y, \lambda_z$	Heat conductivities of material in directions x, y and z, respectively
λ	Heat conductivity of material (for isotropic solid)
M	Movable primitive
M_{id}	Mapping function identifier
$N(\prod)$	Number of nodes in the direction P or V in the FEM mesh of primitive $\prod$
$N(x, y, z)$	Shape function
P	Direction of motion of the primitive M against F
pr	An instance of a primitive

π	*Project* operator
q_0	Heat flux across S_2
$\dot{Q}$	Rate of heat generated in a solid per unit time per unit volume
R^3	Real object space
RP^3	Projective space
ρ	Specific gravity of material
$S_{1, 2, 3, 4}$	Boundary surfaces of heat effects
S^F, S^M	Face labels of the primitives F and M, respectively
σ	The *Stefan–Boltzmann* constant
t	Time
T	Temperature distribution
$\dot{T}$	Incremental rate of T
$\bar{T}$	Prescribed temperature on S_1
T_c	Surrounding temperature
T_r	Temperature of radiation source
θ	Temperature
$\{\theta(t)\}$	Nodal temperature vector at time t
V	Direction perpendicular to both the directions P and D
V^F, V^M	Sets of nodes of the primitives F and M on contacting plane, respectively
(x, y, z)	Coordinates in the real object space R^3
(ξ, η, ς)	Coordinates in the projective space RP^3
^+V	Set of nodes with interpolated analytical data
$*$	*Join* operator
$\ominus$	Set operator for subtraction
ε	Rate of heat radiation
$\{\ \}$	Vector
$[\ ]$	Matrix
$[\ ]^{-1}$	Inverse matrix of $[\ ]$
$[\ ]^e, \{\}^e$	Matrix and vector of element, respectively
$[\ ]^T, \{\}^T$	Transpositional forms of $[\ ]$ and $\{\ \}$, respectively
$|_i$ or $|_j$	Subscript for value of ith or jth node
$|^{new}$	Superscript for values after relative motion took place
$|^{old}$	Superscript for values before relative motion takes place
$|_\omega$	Subscript for values corresponding to x, y and z terms, respectively
$|_P$	Subscript for direction of relative motion
$|_{\bar{P}}$	Subscript for direction perpendicular to P

Abbreviations

2D	Two-Dimensional
3D	Three-Dimensional
AI	Artificial Intelligence
APT	Automatic Programming Tool
B-reps	Boundary Representation
BEM	Boundary Element Method
CAA	Computer-Aided Analysis
CAD	Computer-Aided Design
CAE	Computer-Aided Engineering
CAM	Computer-Aided Manufacturing
CBC	Coded Box Cell
CIM	Computer Integrated Manufacturing
CNC	Computer Numerical Control
CSG	Constructive Solid Geometry
DBMS	Database Management System
DNC	Direct Numerical Control
EFGM	Element Free Galerkin Method
F-label	Face-label
FEA	Finite Element Analysis
FEM	Finite Element Method
FMS	Flexible Manufacturing System
FPM	Finite Point Method
ID	Identification
LAN	Local Area Network
MIT	Massachusetts Institute of Technology
N-label	Node-label
NC	Numerical Control
PC	Personal Computer
RKPM	Reproducing Kernel Particle Method
SE	Spatial Enumeration

Chapter 1
Introduction

1.1 Historical Background

The computer-aided technologies, such as computer-aided design (CAD), computer-aided manufacturing (CAM) and computer-aided engineering (CAE), have a history of more than five decades (from 1958 onward) [1–7].

The first period was mainly dominated by varying military research projects. The development of *Sketchpad* was considered the first exciting demonstration of possibilities in the civilian field [2]. For practical reasons of cost-performance of hardware at the time, initial developments were confined to a limited number of universities and research teams.

The second period ushered in an explosive growth in CAD. A new wave of independent software firms, using minicomputers with better hardware cost-performance, has pushed the development of interactive systems using visual displays leading to graphic workstations. The applications of such workstations interfaced with mainframes or host computers to take advantages of previous investments in software, promoted multiseat configurations and developments in local area networks (LAN) using common databases. In terms of software innovation, there were improvements in operating systems, language and graphics tools, with interfaces to link analysis and application packages. Modular systems were also adopted. Integrated design-to-manufacturing systems emerged for the electronics and microelectrics industries, as well as for manufacturing using numerical control (NC) machines.

The third period was characterised by the emphasis on manufacturing, the use of robots in automated cells for manufacturing and assembly and the integration of all aspects of business. The personal computer (PC) and the major improvements in cost-performance of hardware through volume production and 'chip' technology, also emerged. This, with improved raster and vector displays, enabled the ubiquitous PC platform to be offered as a basic graphic workstation. At the same time the first artificial intelligence (AI) and knowledge-based systems became available for use in many industrial sectors.

L. Wang, *Dynamic Thermal Analysis of Machines in Running State*,
DOI: 10.1007/978-1-4471-5273-6_1, © Springer-Verlag London 2014

In the field of machine tool designs, Massachusetts Institute of Technology (MIT) provided the fertile ground. In the early 1950s, the Computer Applications Group of Electronic System Laboratory (formerly the Servomechanisms Laboratory) pioneered paper tape control of machine tools leading into the late 1950s to the development of the automatic programming tool (APT) language for programming of cutter movements [6]. In 1958, D. Baumann and S. A. Coons of MIT Mechanical Engineering approached D. Ross of MIT Computer Applications Group to see whether it might be possible to take another important step beyond APT [8]. At the Baumann-Coons-Ross meeting, a new system was outlined that would bind man and computer in an intimate co-operative complex, a combination that would use the creative and imaginative powers of humans and the analytical and computational powers of computers, each with the greatest possible economy and efficiency. The outcome of this meeting was a formal arrangement for Computer Applications and Mechanical Engineering to work together in a broad study of what they then named CAD [9].

In the early years, there was some controversy over the nature and characteristics of CAD. For example, did the 'A' stand for automated or aided? Initially, attempts were made to automate the design process by creating a program to duplicate the logical and numerical steps taken by human designers. This approach had many drawbacks, not least the need with batch computing to rerun the program for changes in assumptions or parameters. The more serious objection was the inability to interact with the computer and use human judgement. The breakthrough came when it became possible to control the overall sequence of solving a problem, to think and intervene, feeding information in real time into computers. This interactive concept of a partnership between a designer and a computer transformed problem-solving and decision-making, and was completed with the developments of the storage tube and visual display that allowed designers to 'see what they were doing'.

In 1962, four years on from the Baumann-Coons-Ross meeting, Ivan Sutherland's thesis on the uses of a display for design, with constraints, became well known. His system named *Sketchpad* was the first showpiece of CAD. Following *Sketchpad*, a car panel design system developed by General Motors, with IBM, was presented at the 1964 Spring *Joint Computer Conference* (JCC) as the second showpiece.

The early CAD pioneers were motivated by the need to interact with computers in a new kind of partnership. Much was heard of the man-machine interfaces and the need to develop creative ideas using a graphic display for visual communication. Since translating ideas into reality means describing and defining solid 3D objects, a huge amount of effort was devoted to 3D modelling and visualisation [10–22]. 3D models of objects are necessary communication media in design and manufacturing; whereas computer graphics provided the key to using computers in order to input, manipulate and output the 3D models.

Some practical CAD systems, such as *PADL*-2 [23], *Build*-2 [24], *GMSolid* [25] and *GWB* [26], appeared after *Sketchpad* in late 1970s. At the same time, since the object definition involves a richness of product and material description, and in many cases, the generation of vast volume of data, much attention was paid to the nature and structure of databases to hold such information [27–29]. Language, in respect of the arguments concerning compilers that had raged in the 1960s, ceased to be a problem. Instead it was accepted that job-oriented commands and menus would better serve the purpose of engineers.

Wrestling with major problems led different groups down different paths. Europe was preoccupied with CAD and interactive computing. America pursued the path of batch computing with ever larger mainframes, and pioneered NC manufacturing. Later in the 1970s, with the extension of the activities of the European Community into the realms of advanced technology, regular meetings took place headed by France, Germany and the UK, during which a clear consensus emerged on the nature and importance of CAD.

CAD acted as a focal point for a ferment of new ideas that provided the driving force for fundamental changes in the way computers were used. A new philosophy for translating ideas into reality emerged, leading to a convergence of computing, information and communication technologies, together with an integration of activities, skills and disciplines that were formerly separate. The subsequent impact of computers on industry and society took shape and substance as CAD was developed and matured.

Out of this background of CAD came the important interactive graphics developments, the visual display software tools and the workstations as we know them today [30, 31]. This in turn enabled modular application systems to be developed with better economy and cost-performance.

In mechanical industries, various robots were developed and utilised for automating tool and workpiece handlings, machine operation and the assembly and welding of components. The application of robots introduced new requirements for robot recognition, and stimulated further developments in image analysis with consequent need for integration with the established methods of object definition in CAD design and modelling systems. Artificial intelligence, in the form of expert and knowledge-based systems, developed in two forms: as conceptual front ends to CAD systems, to enable 'what if' questions of manufacturability or cost to be asked during the design stage; or as an aid to expand human mental processes and experience (the so-called 'thinking machine'). The notion of hardware and software as a system involving humans and machines emerged, as experiences demonstrated the need to link activities and to integrate what became known as 'islands of automation'. As this need for integration dominated, the term computer integrated manufacturing (CIM) became popular.

The developments in CAM can also be traced back to the development of numerical control of machine tools in the early 1950s. This first application of computers for control of machine tools opened up the potential for optimisation of manufacturing through computer numerical control (CNC) and direct numerical control (DNC), for small batch production, leading to flexible manufacturing systems (FMS). CNC was first introduced in 1977 [6], which made it possible that a machine tool could be connected directly online to either a dedicated microprocessor or, on a shared basis, to the computer display which formed the user–system interface. As for FMS, protocols and procedures that were developed to link the design offices with the manufacturing and assembly operations in the mechanical industries gave a powerful impetus to integration and standards for system and network linking. This demand for integration, coupled with the more general applications of CAD/CAM across industry, made obvious the need to link all the activities of business.

Meanwhile, a new concept of computer-aided engineering (CAE) appeared in the 1980s. CAE could be considered as a team, embracing the related areas of CAD, computer-aided analysis (CAA) and CIM. They are in addition to the many supporting activities, such as planning, management and control of manufacturing plants through either direct or indirect computer interface. CAE is a combination of techniques in which man and machine are blended into a problem-solving team, intimately coupling the best characteristics of each. The result of this combination works better than either man or machine would work alone, and by using a multidisciplinary approach, it offers the advantages of integrated teamwork [32].

CAD/CAM, CAE, FMS, CIM and many other acronyms are simply describing the use of computers in various aspects of design and manufacturing, and represent logical steps in an evolutionary development although in different time-frames. To date, it is interesting to speculate on the way that neural networks or connectivity machines, with appropriate simulation software, could extend the performance of current systems and deal adequately with the accumulated information and experience. The ability to cope with 'fuzzy data' through associative memory, and learn from the stored information rather than catalogue it in a database for search and retrieval, would represent a major advance bringing together design, production, finance and management. In the field of machine tool design and machining by using computer-aided techniques, various methodologies and modelling methods were introduced recently, such as machining feature extraction [33–37], intelligent CAD [38, 39] and form-feature recognition by using neural-network-based techniques [40].

In this new situation, standards for open system architecture, operating system characteristics, graphics and system building tools, data information exchange, interface connections and communication protocols all assume a new importance. Thus, some researches have been carried out in this field [41–44].

Today, almost everyone is aware of the powerful influence of the high-tech based on the computer-aided technologies in our industrial society. But, a few have been aware of the meaningful influence of machine tools in the Industrial Revolution. The main aspects of the Industrial Revolution have been concerned only with *power* (principally the steam engine), new *materials* (mostly steel) and the many types of *production machinery* (principally for textiles). However, only few have considered the technical development without which the steam engine and the machinery could not have been built, the development without which steel would have been of little significance—the *machine tools*. One can hardly say that the existence of machine tools was a sufficient condition for the Industrial Revolution, but we are certain that it was a necessary condition for the development of the industrial society in which we live [45]. It is, therefore, necessary to carry out a thorough research of machine tools to the same meaning.

1.2 Motivation of Dynamic Thermal Analysis

This book focuses on an integrated design and analysis (CAD/CAE) approach for machine tool design, especially from the perspective of understanding the dynamic thermal behaviours of machines. One of the benefits of dynamic thermal analysis is to make sure that a designed machine will function as expected in running state, where numerous heat sources (stationary or mobile) may affect the accuracy and repeatability of the machine. Therefore, an integrated CAD/CAE system should be able to bridge the gap between engineering analysis and solid modelling, and transfer necessary information between the design and analysis processes seamlessly without any human intervention. The system should also provide a solver for dynamic thermal analysis of the machine, e.g. using finite element method (FEM), in order to evaluate the dynamic behaviour of the machine tool when it is put under working conditions. Designing *perfect* machines in running state rather than in still is the motivation of the research work documented in this book.

The machine tool design and development are activities that have closely engaged humans for over a century. It is only during the past few decades that these activities have been perceived as a systematic process capable of comprehension, analysis and improvement, due to the introduction of the computer-aided technologies mentioned earlier. In order to be functional in its lifetime in a production environment, a machine tool must satisfy the following:

- Within permissible limits, a specified accuracy of shape and dimensions of the workpiece produced on the machine tool together with the required surface finish must be obtained consistently and, as far as possible, independently of the skill of the operator.
- In order to be competitive in operation, it must show high technical performance with economical and energy efficiency.

When considering the design of such a metal-cutting machine tool, its elements can be divided into three groups: the structure; the drives for the cutting, feeding and setting movements; and the operating and control devices. In the individual treatment of each of the three main aspects, different criteria and methods concerning the theoretical approaches and experimental investigations have to be applied carefully. In this book, full considerations will be given to the first aspect, i.e. the machine tool structure, both for the solid modelling and for the engineering analysis.

The machine tool structure normally consists of fixed portions (baseplate, bed, columns, workheads, etc.), together with the moving parts that carry the workpieces and cutting tools. The relative motions between the machine components are important from the viewpoint of machining. Apart from the functional requirements of shapes, accessibility, ease of chip removal, etc., the fundamental criteria for the performance and the corresponding dimensional layout of a machine tool structure lie in its static/dynamic and thermal rigidities.

Aiming at the automation of machine tool design, various researches have been carried out [46–50]. However, there still remains a big gap between the required performance and the real performance of a machine because the problem-solving capability of current software technology is still below the desired level. Two significant problems remain to be solved, due to the lack of suitable algorithms. They are:

1. An effective interface for data exchange between solid modelling and engineering analysis; and
2. An approach to providing necessary information and to enabling dynamic analyses in order to simulate the true behaviour of a machine tool in running state under actual working conditions.

Here, the engineering analyses include the analyses of stress, temperature and other mechanical properties of an engineering design.

These remaining problems have come as a surprise to many solid-modelling researchers who predicted, shortly after the conception of solid modelling, that the engineering analysis would soon be automated [51]. Essentially, the reason for this far from happening is that most analysis codes, i.e. those that are based on the finite element method, require that a solid structure be represented as the union of simple polyhedra. Such a representation is called a cell decomposition in the solid modelling literature [52]. In the finite element literature, the polyhedra are called elements and the union of elements is called a mesh. The problem is that the decomposition of a solid into elements, or mesh generation, is surprisingly difficult. This is especially true for 3D objects because the finite element method requires that the number of elements be minimised to reduce the computation time, and be well shaped to avoid erroneous results. The impact of this is that current mesh-generation algorithms, which work well for simple models, often produce meshes on complex models that require a great deal of manual editing, or that cannot be used at all in dynamic contexts. Because of its generality and widespread use, the finite element method will continue to be the dominant form of analysis for some time. Therefore, it is necessary and urgent to develop an effective interface that can change the solid model data into the finite element data without human intervention.

On the other hand, a satisfactory design of a machine tool should be based on the results of dynamic analyses and/or simulations of the machine tool in order to evaluate its dynamic behaviour at the design stage. Since the relative motions take place among the machine components under actual working conditions, it is difficult to carry out such dynamic analyses using conventional mesh generation methods. A new approach capable of coping with such situation becomes necessary.

The methodology documented in this book is to bridge the gap between solid modelling and engineering analysis, so as to establish an integrated CAD/CAE system for machine tool designs. A new method for dynamic finite element mesh generation named *coded box cell* (CBC) substitution approach is proposed in this book as a kernel to enable dynamic meshing and analysis. Solid model data can therefore be exchanged and translated to FEM model data directly and automatically. There is no distinguishable boundary between the two types of data. Especially, the CBC substitution approach can adjust those finite element nodes whenever they are split up on the contacting surfaces where the relative motions occurred, and hence ensure the dynamic analysis being carried out continuously.

1.3 Organisation of the Book

This book is composed of six chapters, focusing on basic concepts, methodology and processing algorithms. Brief accounts of each chapter are summarised as follows.

Chapter 1 provides a brief introduction to the historical background of machine design and analysis, and clarifies why this research is necessary and what benefits it may bring about.

Chapter 2 presents a detailed description of the techniques used for representation of machine tool models. It begins with the classification of contemporary solid modelling methods, such as *constructive solid geometry* (CSG), *boundary representation* (B-reps), *spatial enumeration* (SE), *Cell Decompositions*, *Sweeping* and *Primitive Instancing*. CSG is adopted as the primary representation in this book for the ease of modelling the machine tool structures. As the basis of solid modelling of machine tool models, data representations (both geometrical and topological) are described. A low-level data structure is proposed for establishment of a primitive library with characteristics of machine tool structures.

Chapter 3 reveals a method for developing a product modelling system used in machine tool design. Based on the design process of machine tools briefly summarised at the begining, considerable requirements to be met in the product modelling system are outlined together with the definition of product model. For the purpose of system implementation, a high-level data structure for both the parts and the entire machine is proposed. A case study on kinematic simulations of a vertical machining centre is carried out using the product modelling system for design validation. Finally, the product modelling system is linked to the integrated CAD/CAE system.

Chapter 4 proposes a new method for dynamic finite element mesh generation, which is called the *coded box cell* (CBC) substitution approach in this book. The chapter begins with the classification of conventional mesh generation methods, followed by the introduction of the CBC substitution approach. Hexahedrons are chosen as the mesh elements for the convenience of automatic generation and modification of FEM meshes, since the machine tools are mostly composed of cuboids and box-type primitives. The basic procedure for dynamic FEM mesh generation using this approach is summarised as follows: (1) Generation of initial hexahedral meshes for the individual primitives and (2) Adjustment of meshes on the interfaces between the primitives, so as to obtain an updated FEM model at the same time when its solid model is completed. Subsequently, a full detail of the CBC substitution approach is described. A new data structure of machine tool models is also proposed for its easy utilisation in FEM mesh generation, based on which several practical case studies are carried out.

Chapter 5 demonstrates the application of the CBC substitution approach to finite element analysis. Since thermal stability is a major factor to be considered in machine tool design, thermal analysis is considered extensively in this book. This chapter introduces briefly the procedures of thermal analysis using FEM and a two-block simplified model. Finite element analysis of temperature distribution of the model is presented secondly. Emphasis is given to the interpolation of the intermediate results between two consecutive computation steps for continuous calculation, especially when relative motions take place between the two blocks. Finally, discussion and evaluation between the proposed dynamic analysis method and the conventional analysis methods are given based on the computational results.

Chapter 6 reviews the key points of each chapter and summarises the findings and benefits of dynamic thermal analysis as applied to machine tool design. Also, the limitations and future work along this direction are highlighted.

References

1. P.E. Bézier, A View of CAD/CAM. Comput. Aided Des. **13**(4), 207–209 (1981)
2. A.I. Llewelyn, Review of CAD/CAM. Comput. Aided Des. **21**(5), 297–302 (1989)
3. K. Preiss, Future CAD systems. Comput. Aided Des. **15**(4), 223–227 (1983)
4. C.H. English, Interactive computer-aided technology. Comput. Aided Des. **9**(4), 243–253 (1977)
5. J. Oian, Trends in Scandinavian CAD development. IEEE Comput. Graphics Appl. pp. 51–58 (1982)
6. W.S. Elliott, Computer-aided mechanical engineering: 1958 to 1988. Comput. Aided Des. **21**(5), 275–288 (1989)
7. M.J.E. Cooley, Impact of CAD on the designer and the design function. Comput. Aided Des. **9**(4), 238–242 (1977)
8. W.S. Elliott, Interactive graphical CAD in mechanical engineering design. Comput. Aided Des. **10**(2), 91–100 (1978)
9. S. A. Coons, An outline of the requirements for a computer-aided design system, in *Proceedings of AFISP, Spring Joint Computer Conference*, pp. 299–304, 1963
10. W. Myers, An industrial perspective on solid modeling. IEEE Comput. Graphics Appl. pp. 86–98 (1982)
11. T.L. Kunii, T. Satoh, K. Yamaguchi, Generation of topological boundary representations from octree encoding., IEEE Comput. Graphics Appl. pp. 29–38 (1985)
12. M.S. Casale, E.L. Stanton, An overview of analytic solid modeling. IEEE Comput. Graphics Appl. pp. 45–56 (1985)
13. C.M. Eastman, K. Preiss, A review of solid shape modeling based on integrity verification. Comput. Aided Des. **16**(2), 66–80 (1984)
14. T.C. Woo, A combinatorial analysis of boundary data structure schemata. IEEE Comput. Graphics Appl. pp. 19–27 (1985)
15. M.M.F. Yuen, S.T. Tan, K.M. Yu, Scheme for automatic dimensioning of CSG defined parts. Comput. Aided Des. **20**(3), 151–159 (1988)
16. J. Flaquer, J.L. Rodil, Boolean operations based on the planar polyhedral representation. Comput. Graphics **12**(1), 59–64 (1988)
17. J.R. Woodwark, Eliminating redundant primitives from set-theoretic solid models by a consideration of constituents. IEEE Comput. Graphics Appl. pp. 38–47 (1988)
18. W.F. Bronsvoort, Boundary evaluation and direct display of CSG models. Comput. Aided Des. **20**(7), 416–419 (1988)

19. F. Yamaguchi, T. Tokieda, A solid modeler with a 4×4 determinant processor. IEEE Comput. Graphics Appl. pp. 51–59 (1985)
20. S.E.O. Saeed, A. de Pennington, J.R. Dodsworth, Offsetting in geometric modeling. Comput. Aided Des. **20**(2), 67–74 (1988)
21. R.C. Joshi, H. Darbari, S. Goel, S. Sasikumaran, A hierarchical hex-tree representational technique for solid modeling. Comput. Graphics **12**(2), 235–238 (1988)
22. A. Kela, Hierarchical octree approximations for boundary representation-based geometric models. Comput. Aided Des. **21**(6), 355–362 (1989)
23. C.M. Brown, PADL-2: A technical summary. IEEE Comput. Graphics Appl. pp. 69–84 (1982)
24. R. Hillyard, The build group of solid modelers. IEEE Comput. Graphics Appl. pp. 43–52 (1982)
25. J. W. Boyse, J.E. Gilchrist, GMSolid: Interactive modeling for design and analysis of solids. IEEE Comput. Graphics Appl. pp. 27–40 (1982)
26. M. Mantyla, R. Sulonen, GWB: A solid modeler with Euler operators. IEEE Comput. Graphics Appl. pp. 17–31 (1982)
27. A. Kemper, M. Wallrath, An analysis of geometric modeling in database systems. ACM Comput. Surv. **19**(1), 47–91 (1987)
28. Y.E. Kalay, The hybrid edge: A topological data structure for vertically integrated geometric modeling. Comput. Aided Des. **21**(3), 130–140 (1989)
29. H. Samet, R.E. Webber, Hierarchical data structures and algorithms for computer graphics. IEEE Comput. Graphics Appl. pp. 59–75 (1988)
30. L. Hatfield, B. Herzog, Graphics Software—from techniques to principles. IEEE Comput. Graphics Appl. pp. 59–79 (1982)
31. P. Chen, Y. Sheu, Object-oriented graphics knowledge bases. Comput. Graphics **12**(1), 115–123 (1988)
32. J.K. Krouse, CAD/CAM–bridging the gap from design to production. Mach. Des. pp. 117–125 (1980)
33. D.B. Perng, Z. Chen, R.K. Li, Automatic 3D machining feature extraction from 3D CSG solid input. Comput. Aided Des. **22**(5), 285–295 (1990)
34. X. Dong, W.R. DeVires, M.J. Wozny, Feature-based reasoning in fixture design. Annals of the CIRP **40**(1), pp. 111–114 (1991)
35. B.K. Choi, M.M. Barash, D.C. Anderson, Automatic recognition of machined surfaces from a 3D solid model. Comput. Aided Des. **16**(2), 81–86 (1984)
36. M. Shpitalni, A. Fischer, CSG representation as a basic for extraction of machining features. Annals of the CIRP **40**(1), 157–160 (1991)
37. S. Joshi, T.C. Chang, Graph-based heuristics for recognition of machined features from a 3D solid model. Comput. Aided Des. **20**(2), 58–66 (1988)
38. S. Ohsuga, Toward intelligent CAD systems. Comput. Aided Des. **21**(5), 315–337 (1989)
39. J. Corbett, J.A.J. Woodward, A CAD-integrated 'Knowledge-Based System' for the design of die case components. Annals CIRP **40**(1), 103–105 (1991)
40. S. Prabhakar, M.R. Henderson, Automatic form-feature recognition using neural-network-based techniques on boundary representations of solid models. Comput. Aided Des. **24**(7), 381–393 (1989)
41. K. Klement, H. Nowacki, Exchange of model presentation information between CAD systems. Comput. Graphics **12**(2), 173–180 (1988)
42. A.Y.C. Nee, A.S. Kumar, A framework for an object/rule-based automated fixture design system. Annals CIRP **40**(1), 147–151 (1991)
43. E. Molloy, H. Yang, J. Browne, Design for assembly within concurrent engineering. Annals CIRP **40**(1), 107–110 (1991)
44. I. Bet, U. Gengenbach, The CAD∗I interface for solid model exchange. Comput. Graphics **12**(2), 181–190 (1988)
45. R.S. Woodbury, *Studies in the History of Machine Tools* (The M.I.T Press, Massachusetts, 1972)
46. R.B. Tilove, Extending solid modeling systems for mechanism design and kinematic simulation. IEEE Comput. Graphics Appl. pp. 9–19 (1983)

47. C.C. Thomson, Robot modeling—the tools needed for optimal design and utilization. Comput. Aided Des. **16**(6), 335–337 (1984)

48. K. Kitajima, H. Yoshikawa, HIMADES-1: A hierarchical machine design system based on the structure model for a machine. Comput. Aided Des. **16**(6), 299–307 (1984)

49. S.H. Kim, K. Lee, An sssembly modeling system for dynamic and kinematic analysis. Comput. Aided Des. **21**(1), 2–12 (1989)

50. D.A. Hoeltzel, W.H. Chieng, Knowledge-based approaches for the creative synthesis of mechanisms. Comput. Aided Des. **22**(1), 57–67 (1986)

51. M.S. Casale, J.E. Bobrow, R. Underwood, Trimmed-patch boundary elements: bridging the gap between solid modeling and engineering analysis. Comput. Aided Des. **24**(4), 193–199 (1992)

52. W.D. Compton, *Design and Analysis of Integrated Manufacturing Systems* (National Academy Press, Washington, D.C., 1988), pp. 167–199

Chapter 2
Data Representation of Machine Models

2.1 Introduction

As we seek to understand the role and the capability of computers in design and analysis of machine tools, two overriding trends may be perceived: the computer graphics which is used for describing machine tool models being designed and analysed, and the coming integration of the information in the databases needed to automate design and analysis activities.

Works on computer graphics initiated almost 50 years ago. Serious efforts have been devoted since then to integrate the graphical interaction with the programs for designing physical systems, such as aircrafts, automobiles, buildings, ships as well as machine tools. It has been recognised that the computer is one kind of representational medium in the same way as drawings. Instead of representing the drawings and incorporating their inherent limitations, the computer can directly represent the target of the representation—the 3D physical system being designed and analysed. In such a computer representation, a single model can ideally depict the information that normally requires many drawings, as well as pages of specifications and engineering data.

Although the engineering drawings of machine tools have served us well in the past, current efforts towards the integration between computer-aided design (CAD) and computer-aided engineering (CAE) have shown that they have serious limitations as the means of geometric object description or definition. The problem is essentially that the drawings are understandable by humans but not by computers. Therefore, it is necessary to establish a suitable data representation and a common database of machine tools for both design and analysis.

To integrate CAD and CAE for machine tools, it is evident that industries must replace the traditional carrier of geometrical information—the drawing—with a computer-based information-carrying form (data representation) that is capable of supporting both the design and the analysis. To automate operations in these areas, this form of data representation must contain information that, being complete, consistent and accurate, is capable of supporting the application programs automatically.

L. Wang, *Dynamic Thermal Analysis of Machines in Running State*,
DOI: 10.1007/978-1-4471-5273-6_2, © Springer-Verlag London 2014

To achieve the integration and the automation mentioned above, the solid modelling with suitable data representation seems to be key.

In this chapter, solid modelling methods for three-dimensional objects are first classified, and then the data representations for machine tool models are introduced. Finally, a database of primitives is established based on an established data structure.

2.2 Classification of Solid Modelling Methods

Contemporary geometric modelling systems are concerned primarily with the geometry. They provide means for defining the shapes of components and sometimes allowable shape variations (tolerances), for positioning component representations to define assemblies, for calculating properties (appearance, mass, etc.) and for generating manufacturing-process data such as NC (Numerical Control) programs.

Essentially, the integration between CAD and CAE necessitates the use of some means of object definition other than the drawing. One such means relies on the construction of a common database using known mathematical terms in the representation of the object. Recent CAD systems possess the capability to define objects in three dimensions. This allows a designer to develop a full three-dimensional model of an object in the computer rather than a two-dimensional illustration. The computer can then generate orthogonal views, perspective drawings and close-up details of the object. Developments in this area include three-dimensional wire-frame, surface and solid modelling techniques [1]. Only the solid modelling technique is classified here due to its popularity and wide adoption in today's CAD/CAE systems.

Solid modelling is distinguished by the use of valid and unambiguous representations of solids. A solid model will involve both surface and edge definition of an object and will furthermore embody a recognition of volumetric details.

Two basic approaches that are used most frequently to solid modelling are constructive solid geometry (CSG) and boundary representation (B-reps) [2].

Constructive Solid Geometry This is a very direct and effective way to construct a solid object model. An object is modelled by a collection of primitives and a set of transformations and Boolean operations. The primitives, such as cuboid, cylinder, cone and sphere, etc., are parameterised. The transformations in this method include translation, rotation and scaling, and are used to define the positions, orientations and other features of the primitives. The Boolean operations, union, difference and intersection, are used to combine the primitives. In CSG, the final shape of a component is described and maintained internally by a tree structure of simpler shapes of primitives.

Boundary Representation This approach keeps a list of the faces, edges and vertices of a model together with the topological and adjacency relationships between them. In this case, the topology is used to determine the set of edges which constitute the boundary of a particular face or which meet at a specific vertex. In other words, a solid is represented by a finite number of bounded faces, each of which is represented by a set of directed edges that bound it, and each edge is represented by two vertices. Each vertex is defined by a coordinate triple. Several kinds of regular surfaces can be used as the basic face elements for describing the solid object. These include planar surfaces (polygons) and analytical surfaces, such as cylindrical, conical and spherical surfaces.

Both methods (CSG and B-reps) have their relative advantages and disadvantages. In CSG systems, it is relatively easy to construct a topologically correct and precise solid model from the available library of primitives. It is compact in its storage requirements, but slow at producing images. On the other hand, a B-reps system gives designers more freedom in building complex models but the validity of the models could be destroyed in the process. It is also more expensive on memory. Table 2.1 gives the schematic illustrations of CSG and B-reps.

Four other unambiguous schemes for representing solids often in conjunction with CSG or B-reps schemes can be summarised as follows according to Bin [2], Meier [3] and Compton [4], which are also schematised in Table 2.1.

Spatial Enumeration   A solid is represented (usually approximated) as a union of quasi-disjoint box-shaped cells 'filled with matter'. The cells may be of uniform size or of varying sizes if generated by recursive binary spatial subdivision. Enumerations of the latter type may be organised as logical trees, called quadtrees in two dimensions and octrees in three dimensions.

Cell Decompositions   A solid is again represented as a union of quasi-disjoint cells, but now each cell may have a distinctive shape, provided that it is homeomorphic to a sphere. Triangulations are the simplest form of cell decomposition, and finite-element meshes are the most widely used engineering embodiments.

Sweeping   A solid is represented as the spatial region traversed ('swept-out') by either an area or a solid moving on a spatial trajectory. There are two kinds of simple sweep representations: translational and rotational. The first is defined by a 2D contour and a straight trajectory along which it moves. The second is defined by a 2D generatrix and a rotational axis usually coplanar with the generatrix. Although sweeping is central to modelling motional processes such as machining and robotic assembly, there are many open mathematical and computational questions surrounding it.

Primitive Instancing   This is a formalisation of the family-of-parts concept. A solid is represented as a particular member of a family by supplying appropriate numerical parameters to a family-specific collection of formulas for displaying members of the family, calculating their mass properties and so forth. Primitive instancing does not allow the combining of representations to create new or more complex objects. It is also difficult or even impossible to derive geometrical and topological properties directly from such schemes.

Table 2.1 Classification of well-known solid modelling methods

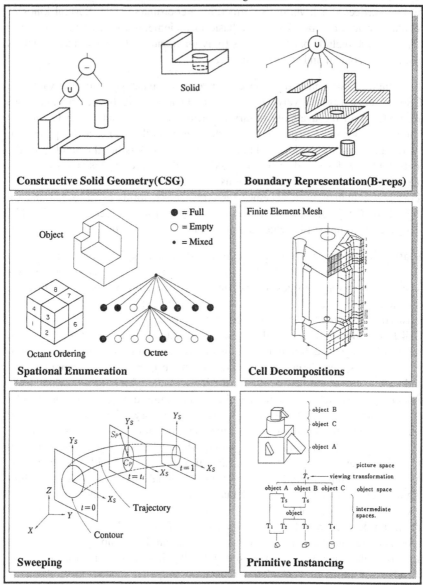

Most commercial solid modelling systems use a hybrid approach, converting the CSG working model into a B-reps definition, and store both models in the computer memory. The CSG might then be used for mass property calculations and the B-reps for downloading edge data for producing 2D drawings.

Table 2.2 Commonly used solid modelling systems

Modeller	Country	Yentor	Representation
CATIA	USA	IBM	B-reps
SHAPES	USA	Draper Lab.	CSG
PADL-1,2	USA	Rochester Univ.	CSG
TIPS	Japan	Hokkaido Univ.	CSG
GDP/GRIN	USA	IBM	CSG
SYNTHAVISION	USA	MAGI	CSG
GMSOLID	USA	General Motors	CSG,Sweep
U.M./BORKIN	USA	Univ. Michigan	CSG,Sweep
GLIDE	USA	Carnegie-Mellon Univ.	CSG,Sweep,EOP
EUCLID	France	Matia Data vision	B-reps
CADD	USA	MCAUTO	B-reps,Sweep
CATSOFT		CATRONIX	CSG
COMPAC	FRG	T.U./Berlin	CSG,Sweep
DESIGN	USA	MDSI	B-reps
BUILD-2	UK	Cambridge Univ.	B-reps
GEOMOD-II		SDRC	B-reps
MEDUSA	UK	CIS Ltd.	CSG,Sweep
UNIS-CAD	FRG	SPERRY UNIVAC	B-reps
GEOMAP	Japan	Univ. of Tokyo	CSG
UNISOLIDS	USA	MCAUTO	CSG
PROREN-2	FRG	Univ. RUHR	CSG,Sweep
SOLIDESIGN		Computer Vision	B-reps
ROMULUS	UK	ShapeData Ltd.	CSG,Sweep,EOP
DDM-SOLIDS		CALMA	B-reps
GEOMED	USA	Stanford Univ.	EOP
ICAD/CAE	Japan	Kobe Univ.	CSG,Sweep

EOP Euler Operations

Several well-known contemporary solid modelling systems are summarised in Table 2.2. *Build-2* and *Glide* are experimental prototypes, and *PADL-1* was designed mainly to demonstrate new algorithms and to serve as an educational and research tool. Most of the other systems in the table are either ready or thought to be nearly ready for industrial use [1]. *Build, Design* and *Romulus* rely internally on boundary representations. They are either created by direct manipulation of such low-level entities as faces and edges, or by means of Boolean and/or sweeping operations. *PADL* systems and *GMSolid* rely primarily on CSG both as an input and as an internal representational medium, but also maintain a boundary representation derived from CSG and therefore guaranteed to be consistent with CSG.

The most important characteristic of solid modelling systems is their ability to support (in principle) any geometric application because they are based on unambiguous (i.e. formally complete) representations.

Finally, the merits and limitations of solid modelling are clarified as follows:

Merits: Solid modelling provides the benefits of designing in 3D, a far more natural mode of expression. It can be used at many different stages in the design and analysis of a machine tool. At the conceptual design stage, it can provide a visual aid, possibly replacing the prototype.

Three-dimensional solid modelling allows sectional views at any position or angle, and exploded views are also possible. The 3D geometry can be transferred for structural stress analysis, dynamic response studies and for the generation of numerically controlled machining data.

Solid modelling also allows for interference checking; design errors which could be catastrophic at the machining or assembly stage of manufacturing can be pre-empted by simulation of the dynamic motions or assembly.

Limitations: In view of the complexities of the geometries generated using the three-dimensional modelling, it usually takes longer to construct a machine tool model in three dimensions than to produce the equivalent 2D views necessary for conventional manufacturing drawings. Furthermore, the computational demands of solid modelling can significantly reduce the performance of the system.

The use of a colour-shaded image is an important facet of the three-dimensional modelling. However, if one requested view does not quite show the details required, then more seconds of computer processing time may be needed before seeking the next view.

The limitations mentioned above can be overcome by a suitable data structure of machine tool models and a primitive library with characteristics of machine tool structures.

2.3 Representation of Machine Tool Models

Machine tool models are represented as combinations of primitives by adopting the CSG modelling technique in this book. Since most parts of machine tools are composed of planar polyhedra and bodies with simple curved surfaces, such as cylinder, cone and sphere, a machine tool-oriented primitive library should be established. Before this work is done, both geometrical and topological representations of geometric components, such as vertex, edge and face, etc., should be discussed first, since individual primitive is defined based on these representations.

Three kinds of geometric components being used here are defined as follows to clarify their constructive roles:

- A *face* is a set of points contiguous in two (not necessarily Euclidean) dimensions.
- An *edge* is a set of points contiguous in one (not necessarily Euclidean) dimension.
- A *vertex* is a bounding point of one or more edges.

Note that a *point* is different from a *vertex*; it is location in space defined by a triple $[X, Y, Z]$. The *space* is considered in its usual Euclidean form as an infinite point set, contiguous in the three dimensions of X, Y and Z.

2.3.1 Geometric Representation

Planar Polyhedra

Classical Cartesian coordinate geometry provides a number of coding schemes for representing the purely geometric aspects of planar polyhedra. Most commonly, a vertex can be represented by a point in coordinate space, a face by a plane equation and an edge by a pair of equations which specify a line, as in Fig. 2.1.

For complex manipulations and transformations in three dimensions, it is common to use a mapping of the objects into the homogeneous representation [5] in four dimensions. The homogeneous representation used in this book provides a more uniform formulation, and allows some extra degrees of freedom that are extremely convenient in numerical computation for dealing with the potential overflow, underflow and the truncation problems that can arise from the degenerate cases. Another advantage of homogeneous coordinates is that the extra degree of freedom in the plane representation allows one to distinguish the inside and outside of the surface. For example, if p is a 4-element column vector representing a plane, then the point represented by the vector V will be on the inside or the outside according to whether $V._p < 0$ or $V._p > 0$.

Topological element	Geometric type	Minimal representation	Number of values	Homogeneous representation	Number of values
vertex	point	(x, y, z)	3	$V = [x\ y\ z\ w]$	4
face	plane	$ax + by + cz + 1 = 0$	3	$P = \begin{bmatrix} a \\ b \\ c \\ k \end{bmatrix} \quad V.p = 0$	4
edge	line	$x = \dfrac{y - y_0}{a} = \dfrac{z - z_0}{b}$	4	$L = \begin{bmatrix} l_{11} & l_{12} & l_{13} & l_{14} \\ l_{21} & l_{22} & l_{23} & l_{24} \end{bmatrix}$ $V = [t\ l]\ .L$	8

Fig. 2.1 Mathematical representation of geometric components

Fig. 2.2 Mapping among
geometric representations

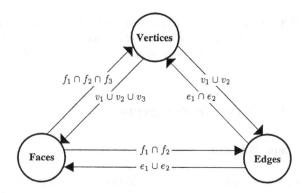

Curved Surfaces

It is always possible to approximate a curved surface using a number of planar faces
to an arbitrary precision. For dynamic thermal analysis, an approximation is quite
sufficient and an exact surface description is not needed. This approximation can be
camouflaged in renderings and displays using various shading algorithms that give
the graphical illusion of curvature without an explicit representation of the exact
surface.

The approximation for curved surfaces is limited to certain classes of simple
surfaces, such as cylinders, cones and spheres, which are considered as necessary
for solid modelling of machine tools in this book.

Mappings Between Geometric Representations

Geometric definitions of the planar face, the edge and the vertex can be derived
directly from each of the others [6]. A vertex is the intersection of two (coplanar)
edges or three faces. An edge is the intersection of two faces or joins of two vertices
(when planar), and a face can be computed to contain two (coplanar) edges or three
vertices, as shown schematically in Fig. 2.2. Similarly, in the case of curved surfaces,
edges and vertices can sometimes be computed from their intersections, with the
restrictions mentioned above. In the planar case, each one of the classes of the
geometric components (vertices, edges and faces) is by itself a complete geometric
representation. The use of all three together is extremely redundant. However, the
minimum size representation is not necessarily the most convenient one and some
redundancy is often justified.

The question of which kind of information should be stored must depend on the
intended use of the system. The solid modelling system presented in this book does
not justify any face, edge and vertex information being stored, but only parameters
from which they may be computed. This will be discussed later.

2.3.2 Topological Representation

Representing the geometric information is only part of the problems of modelling 3D shapes of machine tools. Only if a polyhedron is guaranteed to be convex, will its geometric representation of the faces, the edges and the vertices define a unique topology and thus an unambiguous shape. In the case of concave objects, there is some ambiguity even with oriented planes specifying the inside and the outside. Thus, the other part of the representation problems concerns the question of how the faces, the vertices and the edges are connected.

One approach being used in the solid modelling system is to avoid the issue by limiting the explicit representation to convex shapes. In such a system, the concave objects are represented as a combination of the convex objects, joined at 'virtual' faces. Within the topology of a polyhedron, there are nine classes of relationships among pairs of the three types of the geometric components (vertex, edge and face) [6]. They are shown, with a notation for characterising them, in Fig. 2.3. In notation $\{n_1 : n_2\}$, n_1 indicates a central geometric component, and n_2 a surrounding geometric component to connect with n_1. As with the geometric data, storing all of these relations is highly redundant. In fact, one type of relation is sufficient and all others can be derived from it. However, as some of the derivations are computationally expensive, it is often desirable to store more than one relationship. The question of which to store depends on the application. In the system used in this book, only

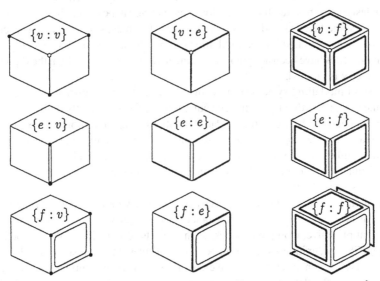

{Central geometric element : Surrounding geometric element}

Fig. 2.3 The nine topological relations among geometric components

relations $\{v : v\}$, $\{e : v\}$ and $\{f : v\}$ are preserved so as to know how vertices are joined and how edges and faces are composed.

2.3.3 Construction of Shapes

Both the geometric and topological representations serve as internal representations of a solid modelling system. To specify shapes of a primitive to the system, only parameters of primitives (defined in detail later) are needed to be entered. These parameters constitute a data structure of the representations, in that the data represented by the parameters is expanded to incorporate the knowledge about the shapes of primitives. Often, there is little relationship between the parameters and the internal data representations since the internal data representations can be considered as a 'black box'. An example might be a cuboid parameterised by width, height and length from which the system will compute a large, sufficient set of shape information based on the internal data representations.

To define a new shape, the simplest and the most common way is as a linear transformation of an existing shape. The transformed shape is often known as an instance of the original, and may consist only of a translation and rotation, or may include scaling and shearing (change in size and proportion). Linear transformations affect the geometry of a primitive but not its topology.

The desire for generality, and for accessing parts of a shape description, has led the system designer to provide the means for constructing shapes from their elementary components: the faces, the edges and the vertices. The means for shape construction reflect the twin aspects of shape modelling—topology and geometry. Broadly, two basic approaches have been adopted, according to the sequence in which the topology and geometry is defined.

The solid modelling system starts with the definition of geometric components, either the vertex coordinates or the face equations, and subsequently define the topological connections between them. Parallel loft lines are interpolated between to define a face. The vertices are formed into rings clockwise to make faces and several faces are combined to form a body. The constructed bodies are checked to see that the faces match up and form a closed surface before a hidden-line removal or a shape operation is carried out.

The methods of shape definition described above may be viewed as hierarchically related, as shown in Fig. 2.4. The instances can be derived by transformation of an original primitive. Complex shapes, such as parts of machine tools, can be built by using CSG technique from a number of primitives and/or instances, while the primitive itself is constructed from the elementary geometric components connected by a topology. The definition of the topology in this system is more primitive than that of the geometry, for reason that it is much more common for many geometries to have a common topology than vice versa.

Fig. 2.4 Hierarchy of definition operations of primitives

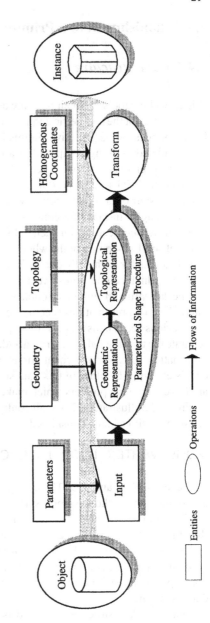

2.4 Establishment of a Primitive Library

2.4.1 Data Structure

When machine tools and their parts are designed by engineers, the elementary design elements are not the geometric components (faces, edges and vertices), but the simple basic volumes called primitives in this book. The machine tools and their parts can be considered as being composed of a limited number of principal primitives. The set of the principal primitives is called a primitive library.

Since the CSG scheme is adopted to represent the machine tool models, it is necessary to establish a suitable primitive library in which each primitive can be chosen as a meaningful instance for constructing a machine tool model. The geometric and topological specifications for each individual primitive in the library are important in terms of solid modelling of the whole structure of the machine tool. Generally, these specifications for each individual primitive can be represented by an appropriate data structure which can best preserve all the necessary information.

The data structure employed could be considered as structural specifications of the internal data representations (geometric and topological representation) discussed in the previous section, as well as in the previous research [7]. Care has been taken to separate the topological structure (face/dege/vertex connectivity) from the geometric information (e.g. surface equations and vertex coordinates) and to keep the one consistent with the other in this data structure. All the relationships between them are linked by using pointers which make the information extraction convenient.

Figure 2.5 illustrates the overall configuration of the data structure adopted. There are seven entities being considered:

EDGE VERTEX LOOP FACE CURVED_F SPLINE_F CONTROL_P

where SPLINE_F and CONTROL_P preserve the attributes needed for B-spline approximations of the free-form surfaces.

The topology of the geometric components within a primitive is schematically illustrated by arrows. The geometries of a primitive are mainly stored in VERTEX, FACE and CONTROL_P. Equations of surfaces stored in FACE make the interference checks among primitives easier and faster. Another benefit of the equations is that approximations of curved surfaces take place at application time, not at model definition time.

Since the data structure proposed is only used for definitions of the primitive library, it is also called a low-level data structure in this book. The data structure for product modelling of machine tools, a high-level data structure, will be introduced in the next chapter.

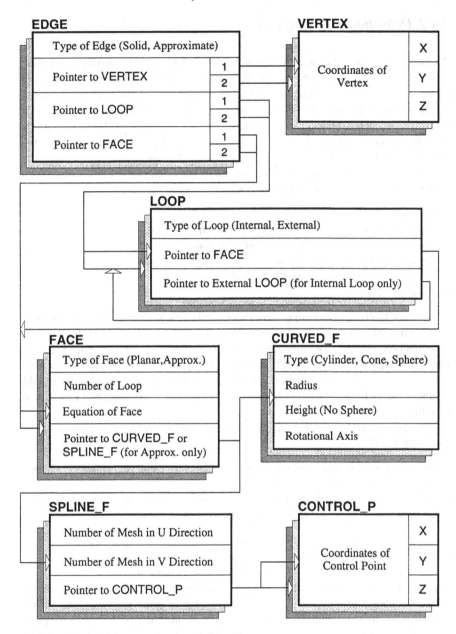

Fig. 2.5 A low-level data structure for primitive library

2.4.2 Definitions of Primitives

As mentioned above, the elementary design elements for solid modelling of machine tools and their parts are not the geometric components, but the primitives. The model of a machine tool is composed of a limited number of principal primitives in the primitive library by applying the CSG scheme. Therefore, definitions of the principal primitives are of vital importance for the solid modelling system. These definitions are carried out based on the data structure introduced previously to organise the topology and to store the geometries for each individual primitive. All the principal primitives in the library, considered in this book, are summarised as follows. Each of them, moreover, may include several variants.

1. Cuboid
2. Cylinder (*Circular Cylinder, Elliplical Cylinder, Prism*)
3. Cone (*Circular Cone, Elliplical Cone, Pyramid*)
4. Sphere (*Sphere, Spheroid, Ellipsoid*)
5. Swept Body (*Translational, Rotational*)
6. Box-type Primitive.

In order to create a certain primitive, one only has to enter a set of parameters which can best represent the primitive. Then, the computer will calculate the other information (geometric and topological) necessary for solid modelling of the primitive according to the low-level data structure. The question here is how to determine these parameters for each individual primitive, since the parameters not only should be able to identify each of the primitives completely but also should be minimum in quantity.

All of these parameters can mainly be divided into two types, in terms of their roles to play. They are:

- The parameters to represent the locations of primitives in Euclidean space, such as (O_x, O_y, O_z); and
- The parameters to represent the geometries of the primitives, such as width L_x, height L_y, length L_z and radius R, etc.

The definitions of the parameters for cuboid, cylinder, cone and sphere are given in Fig. 2.6 together with their schematic illustrations.

The parameters $O(O_x, O_y, O_z)$ in this figure indicate that an origin O of an object coordinates is located in the world coordinates by (O_x, O_y, O_z), representing the location of a primitive. However, only these four types of primitives are not sufficient for solid modelling of the machine tools and their parts. For those parts which have complex cross-sections, the sweep representation seems to be necessary. The sweep representations, both the translational and the rotational, are used in this book. To specify a swept body to the solid modelling system, a 2D cross-section (a contour for translational sweep or a generatrix for rotational sweep) should be defined first. Following the definition for the cross-section, a straight trajectory along which the cross-section moves and a value of sweeping distance are then required

Type of Primitive		Definition of Parameters	Schematic Illustration
Cuboid		Origin: $O(O_x, O_y, O_z)$ Length: L_x, L_y, L_z	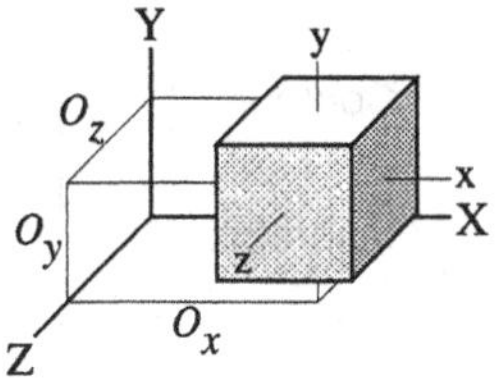
Cylinder	Circular Cylinder $(R_x = R_z = R)$	Origin: $O(O_x, O_y, O_z)$ Radius: R_x, R_z Height: H Side Number: N (for Regular Prism only)	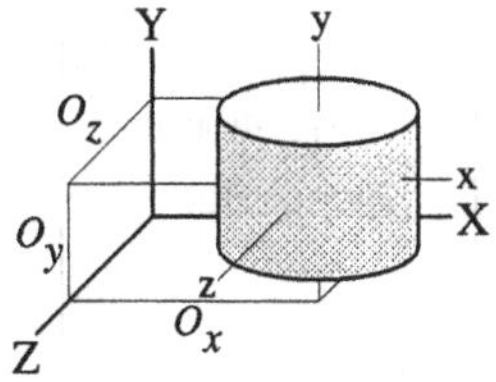 In case of circular cylinder
	Elliplical Cylinder $(R_x \mathrel{!}= R_z)$		
	Regular Prism		
Cone	Circular Cone $(R_x = R_z = R)$	Origin: $O(O_x, O_y, O_z)$ Radius: R_x, R_z Height: H Side Number: N (for Regular Pyramid only)	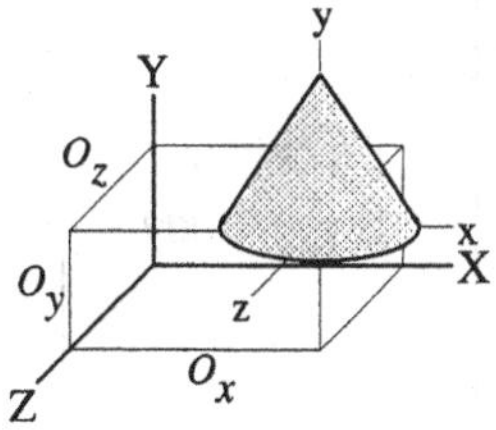 In case of circular cone
	Elliplical Cone $(R_x \mathrel{!}= R_z)$		
	Regular Pyramid		
Sphere	Sphere $(R_x = R_y = R_z = R)$	Origin: $O(O_x, O_y, O_z)$ Radius: R_x, R_y, R_z	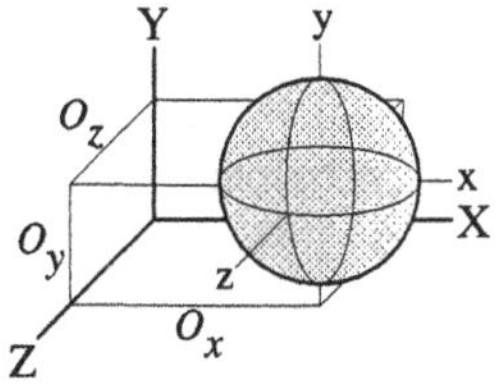 In case of sphere
	Spheroid $(R_x = R_y \mathrel{!}= R_z)$		
	Ellipsoid $(R_x \mathrel{!}= R_y \mathrel{!}= R_z)$		

Fig. 2.6 Definitions of parameters for primitives

for the translational sweep representation. For the rotational sweep, a rotational axis coplanar with the generatrix and an angle are required instead. The definitions for the sweep representations are given in Fig. 2.7. Note that the three constraints shown

		Definition of Parameters	**Schematic Illustration**
Sweep Representation	Contour or Generatrix	Number of Vertex: *NB* *(NB >= 3)*	 *NB = 4* in X-Y plane
	Translational Swept Body	Sweeping Length: *H* *(H > 0)*	*H > 0* along Z axis
	Rotational Swept Body	Sweeping Angle: *alpha* *(0 < alpha <= 360°)* Dividing Number: *Div* *(Div >= 1, within 90°)*	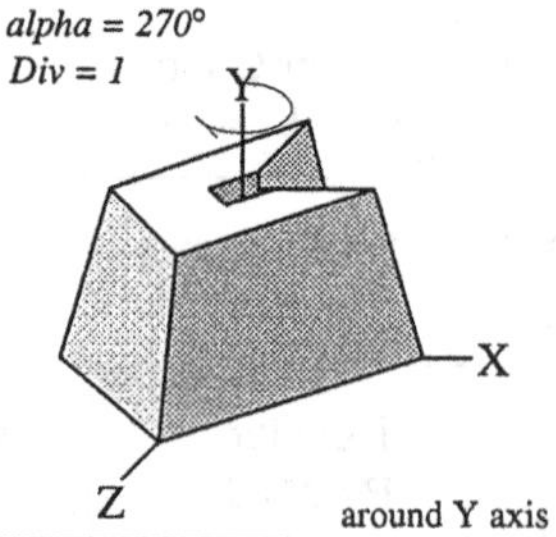 *alpha = 270°* *Div = 1* around Y axis
	Constraints	1. The vertices of the contour (or generatrix) must be given in clockwise-order. 2. The two consecutive sides must not be collinear. 3. The first two consecutive sides can not define a concave part of the contour (or the generatrix).	

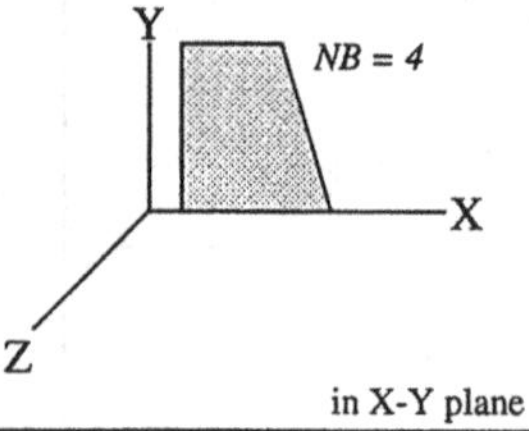

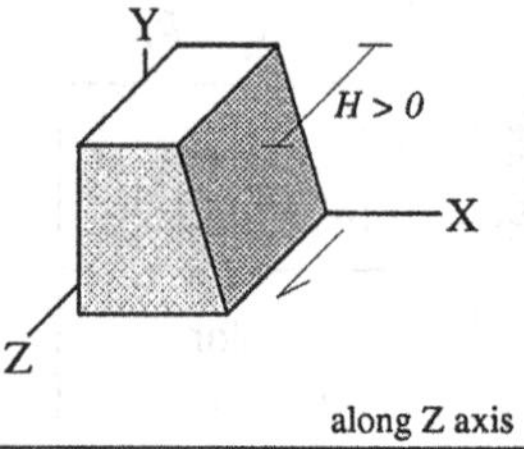

Fig. 2.7 Definitions of parameters for sweep representations

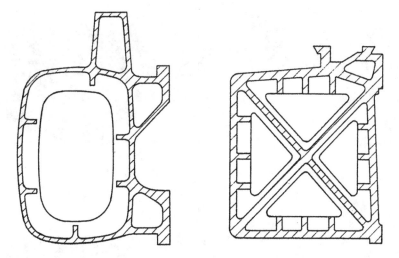

Fig. 2.8 Typical cross-sections of machine tool structures

in this figure should be observed firmly, due to the limitation of the chosen computer graphics.

The primitives defined above are solid, or 'filled with matter'. But most parts of machine tools are not always solid if looking into their internal structures. The layout of the internal structures, as well as the shapes and the sizes of the parts, must be so designed as to ensure the following: (1) the satisfactory conditions exist for the operation and maintenance of the machine tools, (2) the working stresses, deformations, deflections and displacements under working conditions remain within the specified limits and (3) the total weight of the structures and the weight distributions of the parts satisfy the technical and economic requirements. In consideration of these factors, the internal structures of the parts are usually designed in the way shown in Fig. 2.8 to satisfy the requirements of the stiffness, i.e. the resistance to deformation under load, and to minimise the total weight of the machines. However, it should be noticed that the stiffness-to-weight ratio and the actual total weight must be taken into consideration to keep the natural frequency of the machine tool outside the range of its spindle speed in order to prevent any chatter vibration.

Combining these defined primitives properly, the structures of machine tools and their parts can be modelled realistically. This is of importance to increase the accuracy of various analyses and simulations of the machine tool performance, and hence to obtain a much more reasonable structure of the machine tool.

Several samples of the principal primitives in the library are summarised in Fig. 2.9.

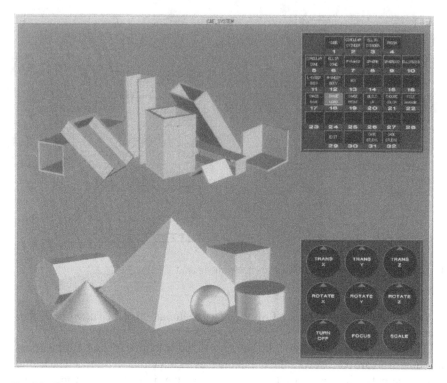

Fig. 2.9   Samples generated from the library

2.5 Concluding Remarks

In this chapter, both the solid modelling methods for three-dimensional objects and the data representing method for machine tool models were discussed. A new data structure was presented, and seven types of principal primitives were defined for the purpose of implementing a primitive library. The main topics discussed in this chapter can be summarised as follows:

1. Two basic approaches that are used most frequently to solid modelling are constructive solid geometry (CSG) and boundary representation (B-reps). The CSG scheme was adopted in this book.
2. Since machine tool models can be represented as the combinations of a small number of principal primitives by using CSG, a primitive library was established based on a specific data structure proposed in this book.
3. Most parts of machine tools are actually designed with thin-walled box sections. Therefore, box-type primitives can be added to the primitive library to provide a realistic way for solid modelling of machine tool structures.

References

1. A.A. Requicha, H.B. Voelcker, A historical summary and contemporary assessment. IEEE Comput. Graph. Appl. **2**, 9–24 (1982)
2. H. Bin, Inputting constructive Solid geometry representations directly from 2D orthographic engineering drawings. Comput. Aided Des. **18**(3), 147–155 (1986)
3. A. Meier, Applying relational database techniques to solid modelling. Comput. Aided Des. **18**(6), 319–326 (1986)
4. W.D. Compton, *Design and Analysis of Integrated Manufacturing Systems* (National Academy Press, Washington, DC, 1988), pp. 167–199
5. E.A. Maxwell, *General Homogeneous Coordinates in Space of Three Dimensions* (University of Cambridge Press, UK, 1951)
6. A. Baer, C. Eastman, M. Henrion, Geometric Modeling: A Survey. Comput. Aided Des. **11**(5), 253–272 (1979)
7. L. Wang, *A Study on Modeling System for Computer-Aided Engineering, Master Thesis, Graduate School of Engineering* (Kobe University, Japan, 1990). (in Japanese)

References

Chapter 3
A Modelling System for Machine Tool Design

3.1 Introduction

In the field of CAE, many systems are now utilised for analysis and simulation of behaviours of machine tools, such as the analysis of stress, strain, vibration and thermal behaviour by applying Finite Element Method (FEM) and/or Boundary Element Method (BEM) and the kinematic analysis of machine products composing of a set of moving components [1–3]. However, these analyses and simulation systems require various modelling techniques to describe the machine parts and components to be analysed, and designers have to generate special-purpose models for individual analysis and simulation from the solid models of the parts and products designed in respective CAD systems. Therefore, effective interface between the solid modelling systems and CAE systems are now recognised as a key technology for realising an integrated CAD/CAE system for machine tool design.

Although some interfaces between CAE and solid modelling systems are utilised in industries, there still remain a number of problems to be solved for further development of effective integrated CAD/CAE system for machine designs. Some fundamental problems are as follows:

1. Development of a product modelling technique which includes both the geometric information and the technological information required for the design and analysis of machine parts and products;
2. Development of useful methodologies for constructing product models of machine parts and products;
3. Development of analysis and simulation systems in order to evaluate the functions and the performances of the designed machine parts and products represented by the product models; and
4. Application of knowledge engineering to the design and analysis of machine parts and products.

This chapter deals mainly with a product modelling system for the analysis and simulation of the machine products composed of a set of moving components, such

L. Wang, *Dynamic Thermal Analysis of Machines in Running State*,
DOI: 10.1007/978-1-4471-5273-6_3, © Springer-Verlag London 2014

as machine tools. The design process of machine products is first analysed, and the requirements for the product modelling system are clarified from the viewpoints of application of the product model to the analysis and simulation. A data structure of the product model is proposed to represent both the geometric and the technological information about the machine products composed of a set of moving parts. An interactive kinematic simulation system is developed to simulate the kinematic behaviour of machine products on the basis of the product model. This chapter is summarised based on the previous research work [4] and the latest extensions.

3.2 Design Process of Machine Tools

3.2.1 General Specification

A machine tool must satisfy the following requirements in order to perform the desired machining operations:

1. Within permissible limits, a specified accuracy of shape and dimensions of the workpiece produced on the machine together with the required surface finish mush be obtained consistently and, as far as possible, independently of the skill of the operator.
2. In order to be competitive in the market, it must show high technical performance with economic efficiency and energy efficiency.

In addition, an intelligent machine tool should possess the abilities as classified here: (1) pre-process, e.g. good interfaces to operators, computer systems and machine control systems; (2) in-process, including online monitoring, self-learning and adaptive decision making; and (3) post-process, such as inspection and maintenance. These functional requirements together with their relations are depicted in Fig. 3.1. It is obvious that these abilities or intelligence cannot be achieved alone through structual design. Instead, the machine intelligence is exhibited in its running state in harmony with other functions as shown in Fig. 3.2, including machine dynamics and control.

The design process of a machine tool may follow different design strategies, depending on the type and purpose of the machine, e.g. general-purpose machine versus special-purpose machine. Nevertheless, the machine to be designed must satisfy the basic motion requirements to reproduce the surfaces of targeted products. This design process can be divided into functional design, structural design and detailed design. Figure 3.3 illustrates the seven steps that a designer should follow, whereas the last three steps are classified as structural and detailed design.

When considering the structural and detailed design of a machine tool, its elements can be further divided into three groups; they are (1) structure, (2) drives for cutting, feeding and setting and (3) operating and control devices. The machine tool structure is primarily considered in this book since it contributes significantly to the thermal behaviour of the machine tool. The machine tool structure generally consists of

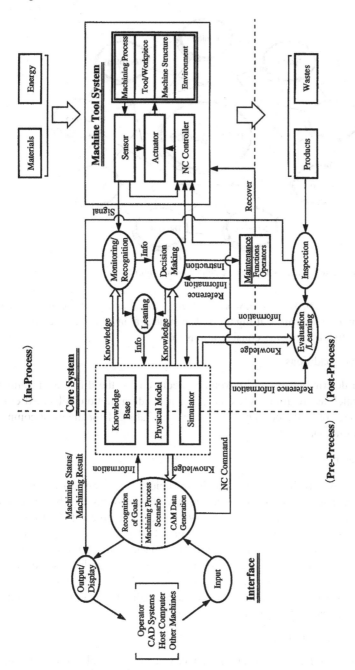

Fig. 3.1 Functional requirements of an intelligent machine tool

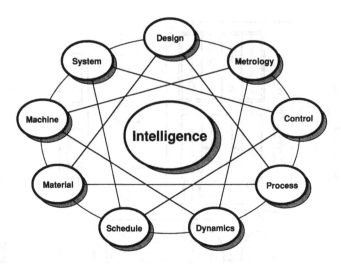

Fig. 3.2 Machine intelligence as a collective behaviour at runtime

fixed portions (base, bed, column, workhead, etc.), together with those moving parts which carry workpieces and cutting tools. The structural layout of a machine tool is determined with the following considerations [5]:

1. Operational conditions

These are determined by the movements required for the different machining processes, which are realised by the cutting and feeding motions allocated either to the workpiece or to the cutting tool as shown in Table 3.1 for the most common machining operations.

(2) Size capacity

This covers not only the overall size of the workpiece which can be accommodated on the machine, but also the overall dimensions which can be covered by the relative motions between the cutting tool and the workpiece. Examples of the former are found in the greatest diameter which is permitted by the swing of a lathe, the size of a casting which can be covered by a boring machine or that which can pass through the gantry of a planing or plano-milling machine. The latter is concerned with details such as the maximum stroke of a planing, milling, shaping or surface grinding machine or the maximum diameter and length which can be machined on a lathe or cylindrical grinding machine.

(3) Performance requirements

These include both quantitative performance (the rate of metal removal, the maximum hole diameter which can be drilled, etc.), and qualitative performance (such as the obtainable degree of accuracy or surface finish).

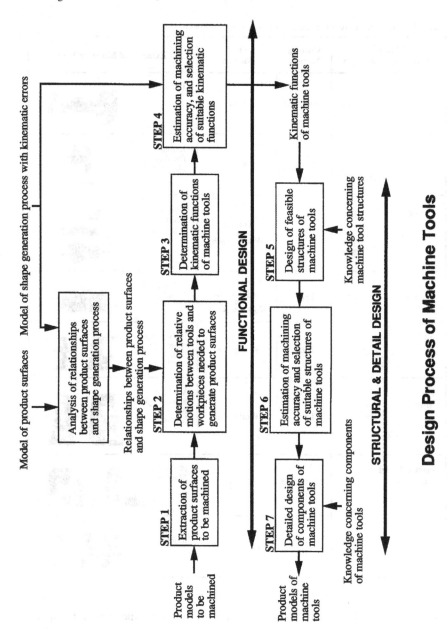

Fig. 3.3 Design process of machine tools

(4) Technical and economic efficiency

In machine structures, the conditions of locating, aligning and guiding different parts are generally determined only by the functional requirements of the operational

Table 3.1 Structural layout of most common machine tools

Type of Machining Operation	Cutting Motion	Feed Motion	Example of Machine Tool
Turning	Workpiece	Tool	Lathe
Drilling	Tool	Tool	Radial Drilling Machine
Boring	Tool	Tool(a) or Workpiece(b)	Horizontal Boring Machine
Cylindrical Grinding	Tool	Workpiece(a) and Tool(b) or Workpiece(a+b)	Cylindrical Grinding Machine
Slab Milling	Tool	Workpiece	Horizontal Knee-Type Milling Machine
Face Milling	Tool	Workpiece	Vertical Knee-Type Milling Machine
Planing(I) and Shaping(II)	Workpiece (I)	Tool(I)	Planing Machine
	Tool (II)	Workpiece(II)	Shaping Machine

movements, which must be related to the applied forces and the operational speeds. However, the design of a machine tool structure must also be concerned with the factors which may affect the productive performance of the machine and which may arise from installation, control and maintenance requirements.

The fixed and movable components which form a machine tool structure must locate, align and guide each other in accordance with the required relative positions between the workpiece and the cutting tool at any instant during the operation of the

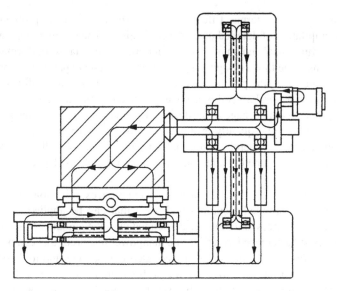

Fig. 3.4 Flow of forces in a horizontal boring and milling machine

machine. They must transmit the weights of the various parts on to their supporting elements and close the flow of the operational forces which are exerted between the workpiece and the tool carrier during machining. The flow of the cutting forces is usually closed within the machine tool structure, as shown in Fig. 3.4 [5] as an example.

The layout of the structure as well as the shapes and the sizes of its components must, therefore, be so designed as to ensure the following: (1) the satisfactory conditions exist for the operation and maintenance of the machine, (2) the working stresses, deformations, deflections and displacements under working conditions remain within the specified limits, (3) the total weight of the structure and the weight distribution of its components satisfy the technical and economic requirements and (4) the efficient manufacturing at competitive cost is possible.

3.2.2 Design Methodology

Design in general terms can be defined as the means by which solutions are contrived to specific problems and in response to a need. This definition draws no distinction between the traditionally separate fields of design, analysis, prototyping and manufacturing. Before examining the several facets of computer-aided engineering for machine tool design, let us consider the general iterative method of design process.

The realisation that a need exists for solving a particular problem will lead to the broad definition of the problem. This will in turn assist the formulation of the problem in engineering terms. Effectively, the formulation of the problem would entail the compilation of detailed design specifications. The specifications will generally include functional and physical characteristics, cost, quality, performance, etc. It is only at this stage that a designer employs intuition and experience to produce a preliminary solution to the problem; a certain type of components, sub-assembly or sub-system of the general assembly is conceptualised by the designer.

The early or conceptual stage of the design process is dominated by the generation of ideas, which are subsequently evaluated against general requirements' criteria. There follows a process whereby additional data are incorporated allowing decisions to be made between competing alternatives as more tangible evidence of function is derived.

An important step in the entire process of design is the examination and allocation of the engineering-oriented resources and tools that can be applied to solve the problem. These resources and tools are associated with the amount of time in which the solution must be achieved as well as the manpower resources that can be devoted to obtain a solution. This step is critical because the resources and tool constraints may affect the approach taken to attack the problem area or even force the scope of the problem to be reconsidered. In both cases, the problem formulation may have to be considered carefully.

In terms of design tools, one interesting observation that has emerged from an earlier survey [6] is that there exist many commercial CAD tools to support detailed designs but few to support conceptual design. The conceptual design can be split into two stages – the first stage in which fuzzy customer requirements are mapped to functional specifications, and the second stage where a design team tries to develop multiple alternative design solutions from functional specifications. There are relatively more tools available to support the second stage of the conceptual design than the first stage (Fig. 3.5). For machine design, the component shape is decided at the second stage before entering in the detailed design.

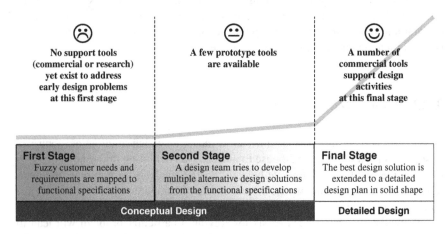

Fig. 3.5 Availability of design tools

The preliminary or conceptual designs are then subjected to the appropriate analysis to determine their suitability for the specified design constraints. If these designs fail to satisfy the constraints, they are then redesigned or modified on the basis of the information gained from the analysis. This iterative process continues until the proposed designs meet the specifications, or until the designer is convinced that the design is not feasible within the specified constraints. The components, sub-assemblies or sub-systems are then synthesised into the final overall system in a similar iterative manner [7].

The above is then followed by an assessment of the design against the specifications established during the problem definition phase. This assessment often requires the fabrication and testing of a prototype model to evaluate the operating performance, quality, reliability, etc.

Clearly, the design process relies entirely on the iterative approach, in the sense that each iteration provides an improvement in the product. The main problem in this iterative approach is that it is time-consuming and many hours of attempts are required before a design project is completed.

Figure 3.6 shows a schematic of the iterative design method. A number of machine tool design tasks can be performed with the assistance of an integrated CAD/CAE system to shorten the design time.

3.2.3 Design Process

Design process of machine tools may be different from one type to another and from one engineer to another. In the design fields of machine tools that require generating and/or transforming kinematic motions, the design process can generally be divided into three steps; they are, conceptual design, fundamental design and detailed design [8, 9].

Figure 3.7 shows the flow of information and process in the conceptual and structural design of machining centres, which are the typical machines generating precise kinematic motions [9]. In this case, requirements on machining centres given by users are first converted to the specifications of kinematic motions of cutting tools and workpieces, and the structural configurations representing rough shapes and dimensions of machining centres are designed. Following this, suitable structural models of machining centres are designed, which satisfy the specifications and structural configurations. The structural models include the information such as geometries of units and connecting relationships among the units. In the detailed design, individual units are divided into parts, and detailed shapes and dimensions of parts are determined. The technological information, such as tolerances, surface roughness and materials are also specified in the detailed design stage taking into consideration the required specifications of both the kinematic motions and manufacturability.

The analysis and simulation are carried out in both the structural and the detailed design stages in order to evaluate the quality and performance of the designed machining centres. During the structural design process, it is examined, through the

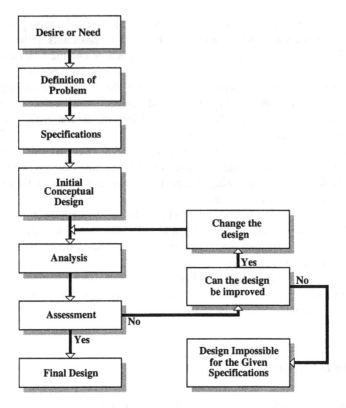

Fig. 3.6 Iterative design methodology

kinematic and thermal simulation, whether or not the designed fundamental structures are suitable for the shape generation specified in the requirements. Various types of analysis and simulations, such as analysis of stress and strain, analysis of deformation, analysis of accuracy and dynamic simulation, are needed during the detailed design process.

Aiming to address the aforementioned challenges in simulation, the remainder of this chapter will discuss a product modelling system for design and analysis of fundamental structures of machines composed of a set of moving parts. The characteristics of machine tools considered here are summarised as follows:

1. The requirements of machine tools are mainly given by the generation and/or transformation of kinematic motions;
2. Machine tools are composed of a set of components;
3. The geometric and the technological information about the physical interfaces between the components are required in the design and analysis of machine tools; and
4. The analysis of kinematic motions and the accuracy of the motions are important in the design verification of the machine tools.

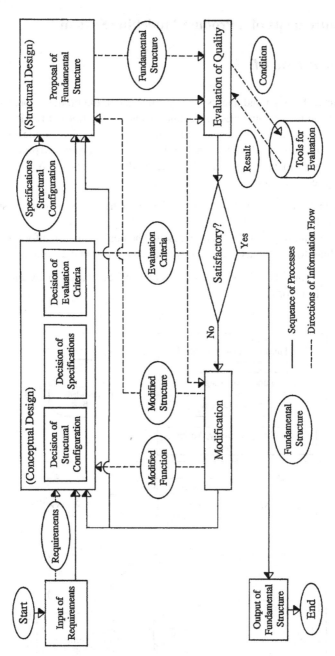

Fig. 3.7 Information flow in conceptual and structural design

3.3 Requirements of a Product Modelling System

3.3.1 Product Models

In the structural design and analysis of machine tools, the models representing the functions, structures, geometries and technological information of machine tools play a key role in connecting the design and analysis processes. The models describing such kind of information about the machine tools are in general called product models. The product information represented by the product models can be classified into two basic types. They are:

- Geometric information which describes the geometries, dimensions and structures of the machine tools; and
- Technological information which represents the non-geometric information about the machine tools such as materials, surface roughness and accuracy, etc.

Figure 3.8 shows the roles of a product model and its relations between the design and analysis functions.

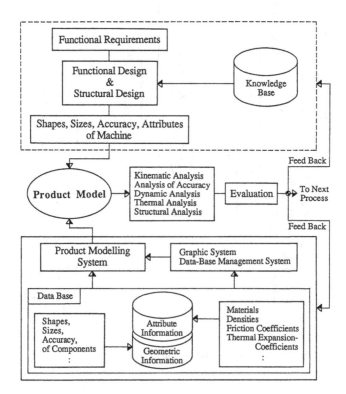

Fig. 3.8 Roles of a product model in design and analysis

The fundamental structures of machine tools are designed in the functional and structural design phases. The shapes, sizes, accuracy and other technological information of the machine tools are determined at this stage. With use of a product modelling system, a product model representing the designed machine tool can be generated based on the geometric and technological information determined during the design processes. The product model is then transferred to analysis and simulation systems to verify the functions and the performance of the designed machine tool. If the designed machine tool does not satisfy the requirements and the specifications, the design processes are repeated and the product models are modified. The detailed designs of the machine tool are carried out based on its product model generated at this stage.

Since the product models are the core models for design and analysis of targeted machine tools, the product models should include all the information about the machine tools required for the design and analysis. The contents of such a product model considered for the design and analysis of a machine tool are summarised below.

1. The functional information such as the requirements and specifications;
2. The geometric and technological information about the individual components of the machine tool. The technological information includes tolerance, accuracy, materials, rigidities, etc.;
3. The geometric and technological information about the physical interfaces between the components, such as fixtures, friction coefficients, etc. The information about the relative motions between the components is also included; and
4. The structural information about the entire machine tool, which includes the connecting relations among the components and the relative placements of the components, etc.

Figure 3.9 provides all the information that a product model may have.

3.3.2 Requirements

Recent technology advancement has encouraged industries to design and produce new types of machine tools with higher quality at lower cost than ever before. The requirements for better performance, greater reliability and lower cost for these products on one hand, and for shorter design/manufacturing time with less manpower on the other hand, are becoming more crucial. Accordingly, the needs for more powerful integrated CAD/CAE systems based on the product models than those currently available are on demand. Since the product models can provide all the needed information during the design and analysis phases, it is urgently required to develop such a product modelling system to support integrated design/analysis activities. The following requirements for appropriate product modelling are taken into consideration in this book:

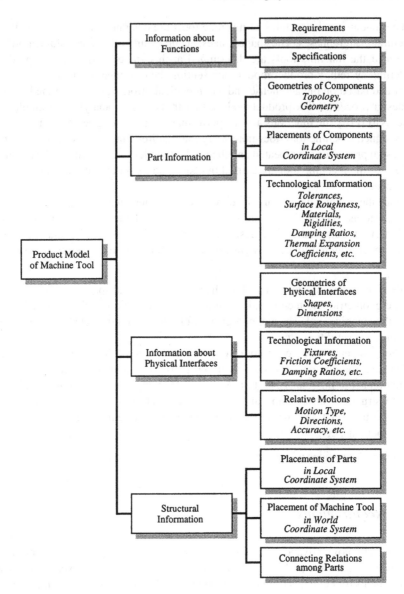

Fig. 3.9 Detailed information contained by a product model

1. The human designer should keep control of the design process but, at the same time, the product modelling system must be able to take part in this process as far as possible.
2. The design is an activity to produce a functional product that satisfies the given requirements. However, in the real-world environment, the requirements are not

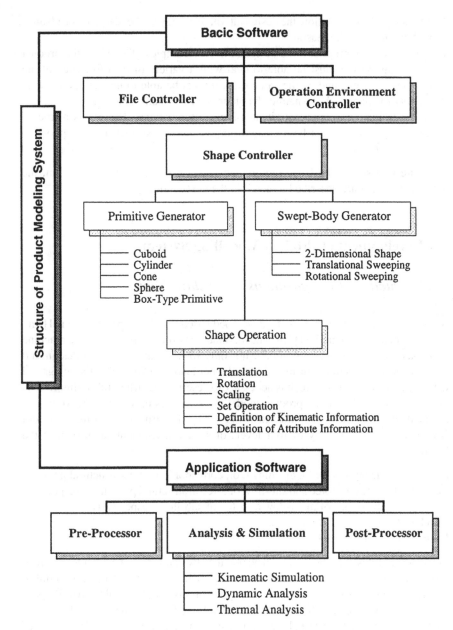

Fig. 3.10 Basic requirements of a product modelling system

always fixed at the beginning but often changed even in the midst of the design process. Therefore, the product modelling system must be able to adapt to the changing situation.

3. To assure the best solution, the design alternatives may not be restricted in advance without any definite reason.
4. Sometimes, the product to be designed is very complex. Therefore, the product modelling system must be able to deal with a variety of data to represent the product precisely, and, at the same time, it must be able to manipulate the data flexibly to adapt to the dynamic change in the design environment.
5. As the design is a complex activity involving different types of problems, a product modelling system must be a general-purpose system so that it can support all aspects of the design.

Figure 3.10 shows the basic requirements of a product modelling system, the outlines of which are described in detail in the next section.

3.4 Development of Product Modelling System

3.4.1 Structural Configurations of Machine Tools

A product modelling system can be developed based on CSG representation [1] and modular construction method [10]. The modular construction for producing machine tools has recently become important, not only to help rationalise the design and manufacturing procedures of machine tools, but also to help evolve the integrated CAD/CAE system. Therefore, it is necessary to describe the structural configurations of machine tools before the product modelling system is introduced. According to the method proposed by Shinno and Ito [10, 11], the structure of a machine tool is composed hierarchically of four levels of structural elements as shown in the following:

Level 1 Whole structure of a machine tool, corresponding to its structural pattern.
Level 2 Main-flow of force and sub-flow of force in structural pattern, corresponding to sub-pattern of main-flow and that of sub-flow, respectively.
Level 3 Pair of structural modules.
Level 4 Structural module.

Since the modules of higher level can be obtained by combining the modules of lower level, the whole structure of a machine tool is created from the structural modules. Figure 3.11 gives the definitions of nine types of structural modules [10]. Pairs of structural modules can be formed in a way that force flows from the initial vertex to the terminal vertex. The vertex here means a certain structural module. The results of possible pairs with use of the nine structural modules as initial vertices are illustrated in Fig. 3.12 [10].

Finally, all the sub-patterns of both main-flow and sub-flow are generated based on the same principle of force flow. The sub-patterns of main-flow use cutting tools as their initial vertices, and those of sub-flow use workpieces as their initial vertices. Both of them own a common terminal vertex, the floor. The whole structures of

	Structural Module	Name of Structural Module		Structural Module	Name of Structural Module
1	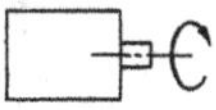	Spindle Head	**5**	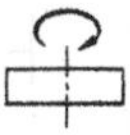	Rotary Table
2		Slide Unit	**6**		Cross-slide Unit
3	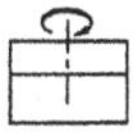	Swivel Slide	**7**		Base (Stationary Unit)
4		Column	**8**		Column Base
			9		Bed (with Longitudinal Slide)

Fig. 3.11 Definitions of structural modules

machine tools, therefore, can be generated by combining the sub-patterns of main-flow with those of sub-flow. The final results [10] of the structural configurations of machine tools are summarised in Fig. 3.13.

3.4.2 Modelling of Part Information

In this book, the CSG technique is adopted for describing part (or machine component) geometries, since most parts of machine tools have relatively simple shapes and the amount of geometric data in CSG representation is smaller than that in B-reps. The CSG technique is also suitable for modification of the part geometries.

Figure 3.14 illustrates the classification of the principal primitives and the necessary parameters prepared in the primitive library. The geometric data of the individual principal primitives are input to the product modelling system and stored in the data structure shown in Fig. 3.15 to form an instance. As shown in the figure, each instance has a list of topological elements, such as the faces and the vertices, together with the topological relationships among these elements. It should be noticed that the data structure shown here is conceptually different from the low-level data structure mentioned in Chap. 2 in terms of: (1) the low-level data structure is used only for definitions of the principal primitives, while the data structure here is applied to generate sets of instances; and (2) since the faces and the vertices of the instances are required for interference check in kinematic simulation, they are therefore included in the data structure. The geometries of parts are thus generated by combining the instances properly using the CSG technique.

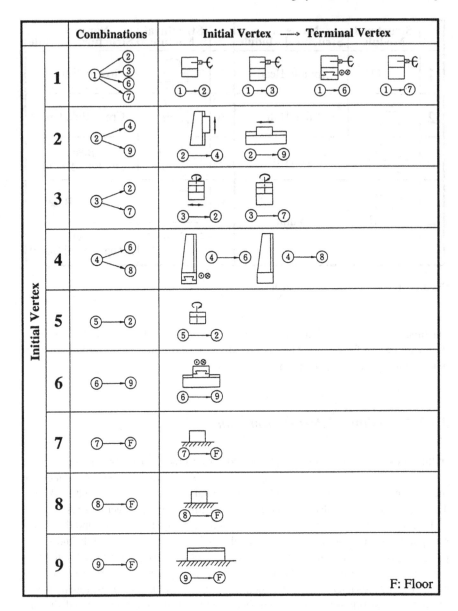

Fig. 3.12 Possible pairs of structural modules

The technological information about the parts is then added to the geometric information shown in Fig. 3.15. The relationships between the technological information and topological elements (faces and vertices) are described by means of pointers (not shown in the figure).

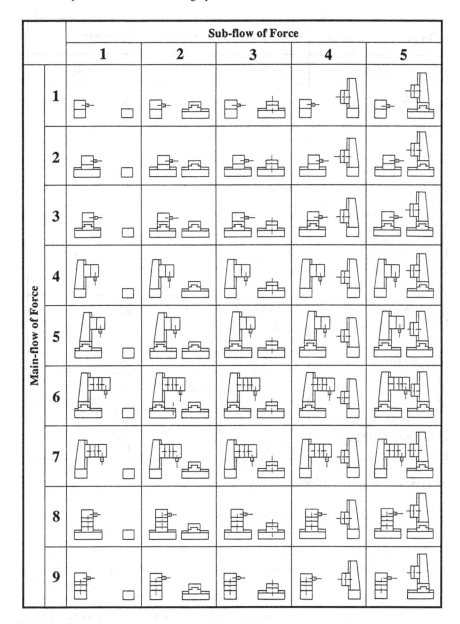

Fig. 3.13 Final results of structural configurations of machine tools

Type of Primitives	Parameters to Describe Primitives	
Cube	Length of Edges	L_x , L_y , L_z
Cylinder Circular Cylinder Elliplical Cylinder Prism	Radiuses Height Number of Side Faces	R_x , R_y H *N(for prism only)*
Cone Circular Cone Elliplical Cone Pyramid	Radiuses Height Number of Side Faces	R_x , R_y H *N(for pyramid only)*
Sphere Circular Sphere Spheroid Ellipsoid	Radiuses	R_x , R_y , R_z
Linear Swept Body	Number of Verties of 2D Shape Sweeping Length	*NV* *SL*
Rotational Swept Body	Number of Verties of 2D Shape Rotating Angle Step Number of Rotation	*NV* *alpha* *Div*
Box Type Constructure	Type Identifier Length of Edges Thickness	*Type_ID* L_x , L_y , L_z T

Fig. 3.14 Classification of defined primitives

3.4.3 Modelling of Structural Information

The information about the machine tool structures is represented with use of graphs in which the nodes show the individual parts and the arcs correspond to the physical interfaces between the parts on which two parts contact with each other. There are two types of graphs: network that includes closed loop(s), and tree that does not include any closed loop. In this book, tree structure is adopted to represent the product models of machine tools. In a graph representing a machine tool structure, the root node of the graph corresponds to the part which is fixed on the floor or the world coordinate system. The parts composing the machine tool are classified into two types. They are,

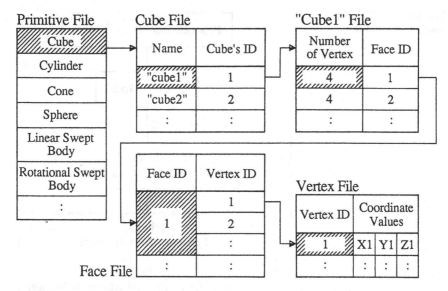

Fig. 3.15   Data structure for generating instance

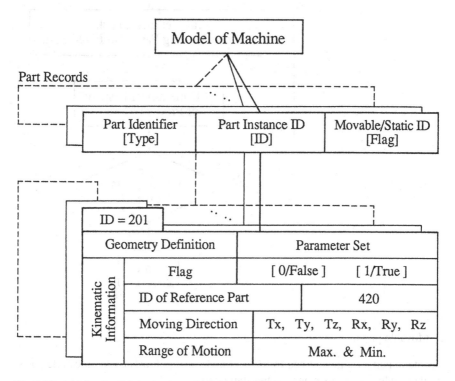

Fig. 3.16   A higher-level data structure of machine tools

Fig. 3.17 A flowchart of
kinematic simulation

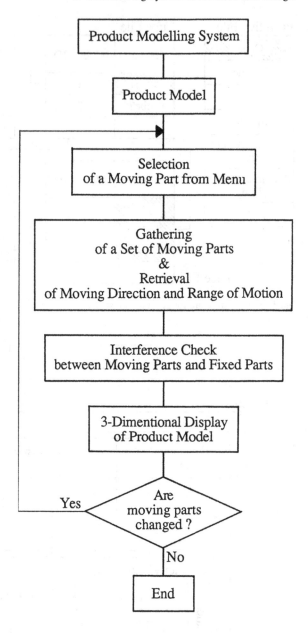

1. Static parts and
2. Movable parts.

The static parts are those fixed to the root node or the floor, and the movable parts
are those which can move against the static parts.

The information about the machine tool structures is described by a high-level data structure as shown in Fig. 3.16. In this figure, the part record is a header of each part which has a part ID, part instance ID and movable/static ID. The part ID and the part instance ID are in fact the pointers to the part models, and the movable/static ID indicates whether or not the part is movable. The detailed information about the machine tool structures and the physical interfaces is given in the lower-level data records.

3.4.4 Kinematic Simulation

The ultimate goal of a machine tool builder is to design and produce machine tools that fit well to the market requirements. This means to satisfy the customer's needs with respect to performance, speed, number of axes, reliability, size capacity, economic efficiency and payload, etc. The machine tool designers are therefore presented with specifications which they have to meet. In order to satisfy the physical requirements without prejudicing the economic requirements, it is essential that the response of the design to input parameters/reference signals be fully understood and this would be best achieved by means of full kinematic and dynamic simulations. Simulation would allow the optimisation of a design option at an early stage, which would in turn mean a more reliable product, and a product that would more closely meet the original specifications.

An interactive kinematic simulation system of machine tools is therefore needed to simulate kinematic motions of parts of the machine tools. Figure 3.17 shows the flow of information processing for kinematic simulation based on the product model. One of the movable parts in the product model is initially selected by the designer. Other parts, either fixed on or driven by the selected part, are retrieved in the second step, and the relative information about the kinematic motions of the parts, such as direction, possible motion range and joint type, etc., is retrieved from the product model. In the third step, kinematic motions of the parts are simulated and the solid model of the moving machine is visualised on the computer screen. This allows the designer to detect any collisions among machine components under varying constraints, intuitively, as if the machine is in the actual running state.

3.5 Case Studies

The results of several case studies are presented here to showcase the capabilities of the integrated product modelling and kinematic simulation system.

Figure 3.18 shows an example of a product model which represents a vertical machining centre. As shown in the graph of Fig. 3.18a, the machining centre consists of four static parts and five movable parts. The product model is shown in Fig. 3.18b. During the kinematic simulation, a table and a spindle head are selected as the

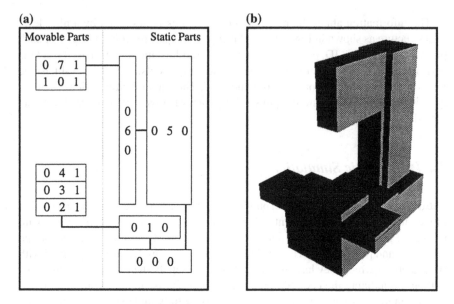

(a)

Movable Parts							Static Parts

```
0  7  1
1  0  1
```

```
0
6    0  5  0
0
```

```
0  4  1
0  3  1
0  2  1
```

```
0  1  0
```

```
0  0  0
```

(b)

Fig. 3.18 An example of a vertical machining centre

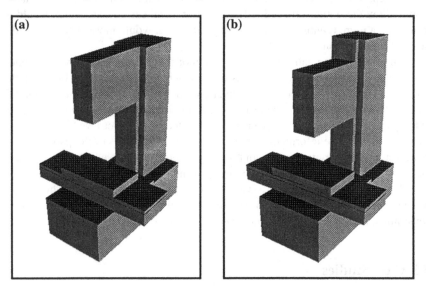

(a)

(b)

Fig. 3.19 Kinematic motions of the machining centre

movable parts and their kinematic motions are calculated and updated as shown in
Fig. 3.19.

According to Ito's research group [11], the structural configurations of machining
centres can be summarised as shown in Fig. 3.20. Several types of machining centres
are therefore taken into consideration in the product modelling system based on

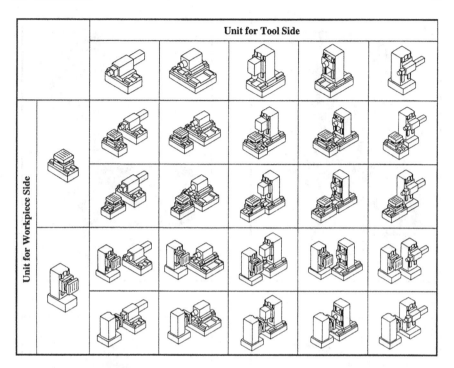

Fig. 3.20 Structural configurations of machining centres

the structural configurations. Figure 3.21 shows the product models of four types of machining centres, whereas Fig. 3.22 illustrates the product model of a robot to demonstrate the utilisation of the product modelling system to other types of mechanisms.

3.6 Concluding Remarks

A product modelling system is needed for realising an integrated CAD/CAE system for machine tool design. After introducing the design process of machine tools, requirements of the product modelling system were discussed on the basis of effective information processing during design and analysis of machine tools. The structural data and information of the product models were clarified into the information about functions, machine components, the physical interfaces among the components and machine structure.

A modelling method was introduced to represent the geometric and technological information about the parts and components of machine tools. The geometric information is described as a set of principal primitives by applying the CSG modelling technique. The technological information is added to the geometric model as

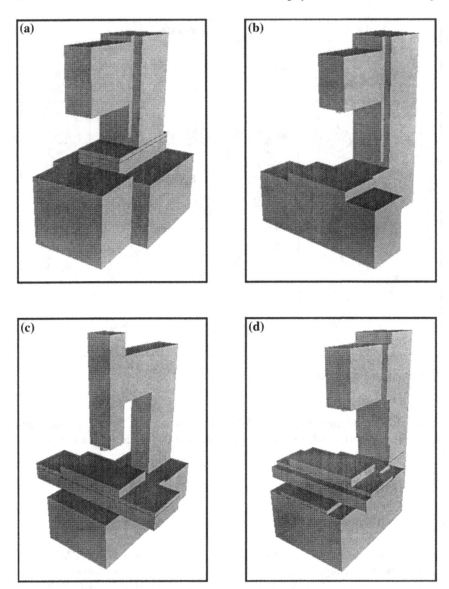

Fig. 3.21 Product models of four types of machining centres

attributes. A high-level data structure was proposed in this chapter. The information about the physical interfaces and the machine tool structures is represented by the high-level data structure, and the information needed for the kinematic simulations is also given in the data structure. An interactive kinematic simulation system was presented to simulate the kinematic motions of moving parts based on the product models of the machine tools.

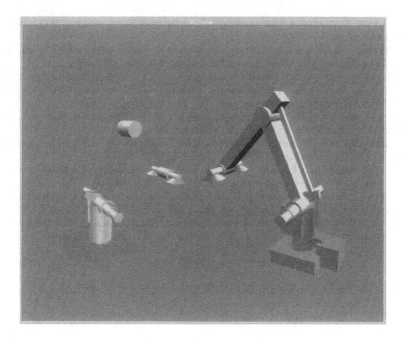

Fig. 3.22 Product model of a robot

A dynamic mesh generation approach will be introduced in the next chapter to facilitate finite element analysis. Its relation and relevance to the integrated CAD/CAE system may be summarised as follows:

1. Generation of initial meshing information for individual primitives.
2. Generation of the solid models of machine tools by CSG technique.
3. Generation of the product models of the machine tools by adding necessary technological information to the solid models.
4. Node adjustment on contacting planes of adjacent primitives while the machine tool structures are under construction.
5. Generation of FEM models by applying further information needed for the finite element analysis.
6. Dynamic node adjustment whenever the relative positions between parts are changed.

References

1. ISO, Industrial Automation-Exchange of Product Model Data. ISO Report, 1990
2. H. B. Voelcker, *Modeling in the Design Process. Design and analysis of Integrated Manufacturing Systems*. (National Academy of, Engineering, 1988) pp. 167

3. C. Mirolo, E. Pagello, A solid modeling system for robot action planning. IEEE Comput. Graph. Appl. **9**, 55–69 (1988)
4. T. Moriwaki, N. Sugimura and L. Wang, Development of solid modelling system for machines with moving parts. Mem. Grad. School, Sci. Technol. **9-A**, 193–206 (1991)
5. F. Koenigsberger, J. Tlusty, *Machine Tool Structures*, vol. 1 (Pergamon Press, Oxford, 1970)
6. L. Wang, W. Shen, H. Xie, J. Neelamkavil, A. Pardasani, Collaborative conceptual design state of the art and future trends. Comput. Aided Des. **34**(13), 981–996 (2002)
7. S. A. Meguid, Integrated Computer-Aided Design of Mechanical Systems, Elsevier Applied Science. (Kluwer Academic Publishers, Dordrecht, 1987)
8. T. Tomiyama and H. Yoshikawa, Extended General Design Theory. in Proceedings IFIP WG 5.2 Working Conference, pp. 95, 1985
9. T. Moriwaki and M. Nunobiki, Knowledge-Based Decision support for Basic Design of Machine Tools. in Proceedings of MSET 21, JSME, pp. 55, 1990
10. H. shinno and Y. Ito, Generation method for structural configuration of machine tools-variant design using directed graph. Trans. JSME. **52**(2), 788–793 (1986). (in Japanese)
11. H. Lee, H. Shinno and Y. Ito, Structural configuration design of machining center-on the variant method using conjunction pattern. J. JSPE. **52**(8), 1393–1398, 1986. (in Japanese)

Chapter 4
Dynamic FEM Mesh Generation

4.1 Introduction

Today, simulation of production processes is becoming much more important in manufacturing, where complex machining operations and materials handling are encountered. A particular emphasis has been given to research and development of virtual manufacturing, having various computer-aided software tools for analysis and simulation of machine behaviours, aiming at realising an optimal production environment, first-time-right products, yet with high quality, low cost and short lead-time. It often requires an advanced system capability to analyse and simulate the dynamic behaviours of production cells and lines, as if they are under real operating conditions. This type of simulation considers production rate of the entire system while treating each machine as a black box. In order to evaluate and optimise the mechanical integrity and to ensure the true dynamic and thermal behaviours of machine tools, some types of analyses are required for individual machines in addition to system-level simulation. Due to the complex nature of the geometric features of the components and that of the applied loads, finite element method (FEM) and boundary element method (BEM) have been mostly adopted in the last three decades. FEM is a powerful numerical tool for solving mathematical problems related to practical engineering situations. In the past, it was a common practice to over-simplify such problems to the point where an analytical solution could be obtained. Because of the uncertainties associated with such a procedure, large safety factors were introduced in the design of machine tools.

FEM has advanced from a numerical procedure for solving the structural problems to a general numerical procedure for solving a differential equation or a system of differential equations. This advancement has been assisted by the development of high-speed and high-performance computers. In the last decades, a large number of CAE systems have been developed and put into practical use in product development, such as analyses of stress, strain, vibration, etc. Besides, there is a growing interest in the development of meshless methods for numerical solutions of partial differential equations. Some typical meshless methods include *reproducing kernel*

L. Wang, *Dynamic Thermal Analysis of Machines in Running State*,
DOI: 10.1007/978-1-4471-5273-6_4, © Springer-Verlag London 2014

particle method (RKPM) [1], *element free Galerkin method* (EFGM) [2], *finite point method* (FPM) [3] and others [4–8]. However, considering the accuracy and stability of the analytical results, FEM remains the most popular and practical approach available today for solving various complex engineering problems. Unlike the meshless methods, FEM inherently requires a properly meshed geometry before any analysis can take place. In other words, all FEMs involve dividing the physical system into some small sub-domains known as elements. Each element is essentially a simple unit, the behaviour of which can be readily analysed. The features of the overall system are accommodated by using a large number of elements. Indeed, one of the attractions of the finite element method is the ease with which it can be applied to real engineering problems involving complex geometrical features. The price that must be paid for flexibility and simplicity of individual elements is in the amount of numerical computations required to solve the resulting sets of simultaneous algebraic equations.

The FEM analysis requires various modelling techniques to describe the components of the machine tools to be analysed, and the designers have to generate FEM models (finite element meshes in most cases) for the analysis from their solid models and to transfer the necessary information from the solid models to the FEM models. Although FEM is a powerful and versatile analytical tool, its usefulness is hampered by the need to generate such kind of meshes. Most of the finite element calculations for generating meshes involve a great deal of effort to generate numerical data such as node numbers, nodal point coordinates, parameter values and element connectivity. This work can be very time-consuming and error-prone if done manually, especially when a large number of elements or very complex geometries are encountered. In recognition of this problem, a large number of methods have been developed to automate the mesh generation task [9–64], putting as few constraints on the user as possible. However, these conventional mesh generation methods are not suitable for the case of dynamic analysis of the machine tools that are in the running state.

Iterative analysis usually requires the mesh being partially refined or even regenerated to increase analytical accuracy. Despite a large number of meshing methods developed in the past, algorithms for dynamic mesh generation and localised mesh adjustment are still missing from the literature. There is still no fully automated mesh generation system for 3D models [65] that are changing dynamically in relative positions. Therefore, it is urgently required to establish a systematic approach that not only can handle the re-meshing problem effectively, but also can simulate a machine's behaviour dynamically and accurately, especially under real operating conditions. A complete understanding of a machine's dynamics or its real-life behaviour is vital to both the structural design and to the run-time machine control.

In this book, a new mesh generation method called *CBC substitution approach* is introduced to meet the need for dynamic FEM mesh generation, and an integrated CAD/CAE system is developed to fully utilise the newly introduced method. Figure 4.1 illustrates the overall configuration of the integrated CAD/CAE system, extended from the product modelling system of Chap. 3. The solid model data of machine tools are first generated in the product modelling system to represent the structural and technological information as well as the geometrical information about

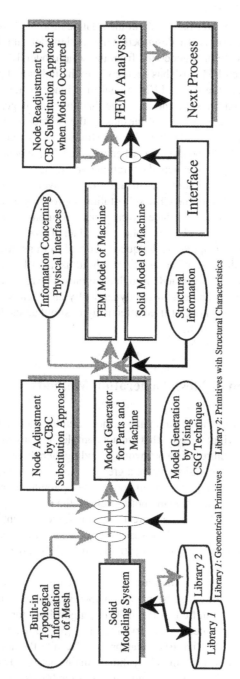

Fig. 4.1 Overall system configuration

the machine tools. The information about the relative motions between the components of the machine tools is also described in the solid model data. The FEM models of the machine tools are then generated from the solid model data. Since the initial information about the FEM mesh has been given to the individual primitives at their solid modelling stage, the whole FEM model of a machine tool can be obtained at the same time when the solid model of the machine tool is generated. It should be noted that the node adjustments of meshes on the contacting planes of the adjacent primitives are carried out by the CBC substitution approach. The FEM model data generated should only be modified whenever the relative positions between the components are changed under actual operating conditions.

Following a brief review and classification of the available meshing methods, the processing procedures of the CBC substitution approach are described in detail. A systematic method is then introduced to carry out the automatic and dynamic finite element mesh generation based on the solid model data of machine tools discussed in the previous chapter. A mapping technique is adopted to transform curved surfaces to planary ones for the ease of automatic mesh adjustment by applying the same algorithm. The algorithm is verified through examples. The results of the case study show that a FEM mesh can be adjusted dynamically and locally around its border zone; and the algorithm can be utilised effectively to simulate its thermal behaviour under real operating conditions.

4.2 Classification of Mesh Generation Methods

The mesh generation methods, automatic or not, are classified in this section. The methods are still evolving; the ones that are not completely automatic at the moment may become so in the future. It is worthwhile including them here. As mentioned earlier, most mesh generation tasks involve a great deal of effort to generate numerical data, such as node numbers, nodal point coordinates, parameter values and element connectivity. Besides, there are many bottlenecks in the meshing process: geometry cleanup, geometry simplification or de-featuring and finite element mesh generation. During the last three decades, many research projects have been carried out to automate the FEM mesh generation process [66], based on the geometric models created by using a CAD system.

Initial efforts in the meshing literature were focused on the two-dimensional mesh generation algorithms based on triangulation [30–32] and quadrilateral elements [32, 67]. Attention then shifted to the meshing algorithms for three-dimensional geometry [33–35] and geometry with curved surfaces [35–39] using tetrahedra, hexahedra and other polyhedra like triangular prisms. Despite the accomplishments in the past, research on new meshing algorithms remains active. Owen et al. [68, 69] used an algorithm called *H-Morph* to generate a hexahedral-dominant finite element mesh for arbitrary volumes. As an extension of the *Q-Morph* algorithm [70], the *H-Morph* method starts with an initial tetrahedral mesh and systematically transforms and combines tetrahedra into hexahedra, without the need of decomposing an arbitrary volume. On the contrary, Lu et al. [71] use a feature-based approach for volume decomposition before applying an appropriate and well-established meshing algorithm to each of the meshable pieces. Li et al. [72] presented a unified scheme for simultaneously refining and coarsening a mesh. This adaptive meshing algorithm is suitable for a domain that has a moving boundary.

Notable efforts have also been given to the automated hexahedral mesh generation, because hexahedral mesh generally provides higher 'artificial stiffness' and better solutions than the tetrahedral one for many physical problems [73]. Sheffer and Bercovier [74] introduced a hexahedral meshing algorithm based on the *embedded Voronoi graph* (EVG), where the EVG is used to decompose a nonlinear volume into simple sub-volumes meshable by basic meshing techniques. Lai et al. [75] proposed an enhanced sweeping mesh generation algorithm by creating all hexahedral element meshes between multiple source surfaces and multiple target surfaces, while most other sweeping techniques [76–78] create all hexahedral element meshes by projecting an existing single-surface mesh along a specified trajectory to a specified single target surface. Dhondt [79, 80] presented a method that can generate an unstructured 20-node brick element mesh for arbitrary structure, based on a triangulation of the structure's surface. Starting from a hexahedral master mesh encompassing the structure, the elements that are intersected by the triangulation are determined, cut and re-meshed according to their cutting topologies. *Whisker Weaving* [81, 82] is another advancing front algorithm for all-hexahedral mesh generation. It uses global information derived from grouping the mesh dual into surfaces to construct the connectivity of the mesh, and then positions the nodes afterwards.

Based on the previous literature surveys [65, 73, 78] and the author's own investigation, major meshing algorithms available in the public domain can be classified and summarised, as shown in Fig. 4.2, in terms of the techniques used and the topologies applied. Most meshing approaches can also be applied to 3D cases.

The five major classes shown in Fig. 4.2 are further explained as follows. One additional minor class of meshing is introduced at the end of this section.

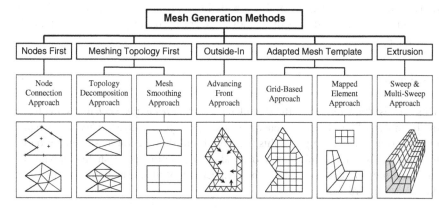

Fig. 4.2 Classification of mesh generation methods

Nodes First

In this class, the nodes are created first in the final mesh, and they are then connected to form triangular or quadrilateral elements. The distribution of the nodes should conform to the mesh density specifications. Then, in the finite element generation phase, one only has to consider how to connect the nodes to form the best possible elements. This approach, which covers the bulk of the mesh generation literature, is called the *node connection approach* [22–30].

This mesh generation approach is very popular because it is conceptually simple. There are two main phases in this approach; they are

1. Node generation (random node generation [22–24], nonrandom node generation [36]) and
2. Element generation [26–29].

Meshing Topology First

This can be divided into two subclasses. If the vertexes of the object are assumed to be the only nodes in the mesh, a triangulation algorithm [15–18] can be applied to create a minimum number of non-overlapping triangles that cover the object. This minimal set of triangles is determined mainly by the topology of the object. The complex topology of the object has been decomposed into the simple topology of the triangles, and therefore this approach is called the *topology decomposition approach* [19–21]. The mesh thus created has distorted and mostly coarse elements, and cannot be used for analysis purposes. Mesh refinement, possibly with element rearrangements by the diagonal transpose technique [15], must be applied to produce reasonable meshes.

The other one is called the *mesh smoothing approach*. The mesh topology is determined first, followed by the nodal positions. Once the mesh topology has been determined, the mesh smoothing techniques [13, 14] are used to calculate the positions of the nodal points. The problem with this approach is that there is no known algorithm for creating the mesh topology, except by using other automatic mesh generation methods.

Outside-In

This class is characterised by the so-called *advancing front approach* that starts from the boundary of an object by generating meshes, layer by layer. *Whisker Weaving* [81, 82] is a typical example of this kind, but for all-hexahedral mesh generation in this case. It uses global information derived from grouping the mesh dual into layers and then into surfaces to construct the connectivity of the mesh, before positioning the nodes forward.

Adapted Mesh Template

A mesh template is pregenerated elsewhere, and then adapted to the object being meshed. Two main approaches within this class can be identified; they are

1. Grid-based approach [40–47] and
2. Mapped element approach [48–57].

The *mapped element approach* includes the *conformal mapping approach* that is relatively undeveloped compared to the other approaches [65].

Extrusion

As its name suggests, this class of meshing approaches first generates an acceptable layer of 'good' mesh and then propagates the layer along a trajectory by means of sweeping to create a complete meshwork. Examples in this class include (1) *sweep approach* [76–78] that creates all hexahedral element meshes by projecting an existing single-surface mesh along a specified trajectory to a specified single target surface; and (2) *multi-sweep approach* [75] that creates all hexahedral element meshes between multiple source surfaces and multiple target surfaces. Both approaches, however, are constrained by the relatively simple shape geometry of the objects to be meshed.

Nodes and Elements Created Simultaneously

In addition to the aforementioned meshing classes, sometimes the nodes and elements are created simultaneously, with no distinctive phases that can be labelled 'nodes only' or 'elements only'. In this class, an attempt is made to generate good elements by considering the object geometry, in contrast to the topology decomposition approach that mostly ignores the geometry (except for the purpose of interference checking). One approach falling into this class is called the *geometry decomposition approach* [58–61].

There are altogether six classes of mesh generation methods, each of which represents one or more approaches for mesh generation of 2D cases, mainly. There are few published 3D mesh generation methods compared to 2D methods due to the greatly increased complexity. Several mesh generation approaches offer the quadrilateral element as an output in 2D, but none has been extended to offer the hexahedral element for the 3D cases [65]. The following 3D mesh generation methods in the literature are classified here, although they are not completely automated [65]; the 3D mesh generation methods are:

- Topology decomposition approach (3D) [19, 20, 62],
- Node connection approach (3D) [33, 34],
  (the initial nodes need to be created separately)
- Grid-based approach (3D) [63],

- Sweep and multi-sweep approaches (3D) [75–78] and
- Geometry decomposition approach (3D) [64].

The classification in this section aims not only to show a clear overall picture of the mesh generation literature, but also to give insights into the relationships among the varying approaches. Finally, it is necessary to clarify that the approaches are still evolving; and therefore the present limitations in their capability may be overcome in the future.

4.3 Concept of Dynamic Mesh Generation

4.3.1 General Considerations

When a milling machine is boring an engine block, the relative positions among its components (table, column, spindle, etc.) vary with time. It is generally difficult to analyse and simulate the dynamic behaviour of the milling machine using conventional FEA (finite element analysis) tools while its components are moving. This is because most of the FEA tools would require the mesh to be regenerated once the corresponding geometry model is changed, or because the nodes of a meshwork on the boundary surfaces between the parts disagree if the nodes move with the model of the machine components. It becomes impossible if the geometry model moves and changes rapidly due to the nodal disagreement. A systematic method is therefore required to generate the FEM meshes of the machine dynamically, aiming at realising the FEM analysis of the machine under running state, and to automate the FEM mesh generation process. Such kind of dynamic mesh generation does not require the entire mesh to be regenerated, and it considers mesh generation in a constructive way when the geometry model of a machine having movable components is designed. The fundamentals of this concept are illustrated in Fig. 4.3.

Information about the initial FEM meshes is generated and preserved to each primitive at the initial stage of its solid modelling, instead of generating the meshes when the final solid model of a machine tool is completed. Hexahedral elements are preselected in this approach due to the ease of mesh alignment. The final solid model of the machine is represented by assembling the primitives that preserve the information about the initial meshes. The final FEM meshes of the machine are then created automatically from its solid model data, and modified dynamically whenever the relative positions between the machine components are changed. In other words, when the primitives are combined to build the machine, the FEM elements near their border zone are adjusted or modified locally, so that the mesh of one primitive connects with that of another smoothly across the boundary. This operation needs to be performed only once if the geometry model consists of non-movable primitives only. Otherwise, the mesh along the boundary of the movable primitives has to be adjusted on the fly whenever its relative position is changed. This dynamic mesh modification is limited to those finite elements whose nodes disagree with each other within the border zone and the elements being affected.

Fig. 4.3 Concept of dynamic mesh generation

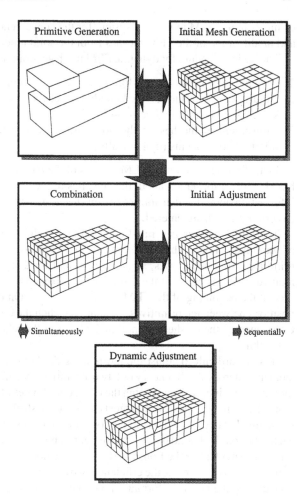

The nodes of mesh elements on the boundary of the two adjacent primitives are adjusted locally by either changing the topology of the elements or by moving their nodes on the boundary surface for mutual matching. Based on this concept, it is possible to generate a FEM model of a machine tool dynamically when the machine is in running state, and to make the meshing process automated, constructive and suitable for modification of both the solid models and their finite element models.

4.3.2 Basic Procedures

The machine tools are generally composed of cuboids and box-type primitives as mentioned in the previous chapters. Therefore, hexahedrons are chosen as the mesh elements in the FEM modelling system for the convenience of automatic generation and modification of the FEM meshes from the solid model data of the machines.

The basic procedure for dynamic FEM mesh generation is summarised as follows:

1. Generation of initial hexahedral meshes for individual primitives;
2. Adjustment of meshes on the boundaries between the primitives constituting the same component (only once); and
3. Adjustment of meshes on the boundaries between the primitives constituting different components (dynamically).

In the first step, the initial FEM meshes of the individual primitives are generated based on the information about the densities of meshes given to the primitives. The information about the densities of the meshes is given to individual primitives when the primitive data are entered.

The initial meshes generated are modified at the boundaries between the pairs of primitives in order that the meshes of one primitive agree with those of another primitive at the boundaries, in the second and third steps. In the second step, the relative positions between the primitives are fixed. Therefore, the meshes are modified at the beginning of the FEM model generation. On the other hand, the relative positions between the primitives are changed when the components move relatively. Consequently, the meshes are modified whenever the relative positions are changed, in the third step.

It is not difficult to adjust the mesh along the boundary if the sizes of the mesh elements (across the boundary) of two primitives are the same. The mesh can be modified, in this case, by adjusting the nodal coordinates of those conflicting elements along the boundary. However, in most cases, the mesh elements of one primitive must be subdivided in order that the numbers of the elements of both primitives agree with each other on the boundary. A requirement to be met in subdividing an initial mesh into some smaller hexahedrons is to localise the modification only in the area near the boundary so as to increase the efficiency of dynamic adjustment. The question here is how to keep the mesh conformation without introducing distored elements and how to minimise the changes in elements of the initial FEM model. Such conformation is generally difficult to achieve automatically because of the unidentified features of the elements, especially for 3D cases [37, 65].

In order to solve the problem, a set of pre-cut hexahedral cells, called CBC (*Coded Box Cells*), are proposed to subdivide the mesh elements into several smaller cells and to increase the number of the mesh elements of the primitives. Figure 4.4 shows the topological structure of the CBCs. The CBCs are defined based on the way that a hexahedral mesh element is subdivided. The first row and the first column of the figure describe how to subdivide an element if viewing it from Z and X directions, respectively. In the figure, the Y-axis coincides the normal vector of the contacting surface (boundary surface) of any two primitives. The X- and Z-axes show the directions along which the hexahedral mesh element is subdivided. Since CBC_{34} is identical to CBC_{43} in their topological structure, only the six CBCs within the bold line in Fig. 4.4 are used for mesh adjustment. If a mesh element is required to be subdivided, it is simply substituted by an appropriate CBC to change its topology.

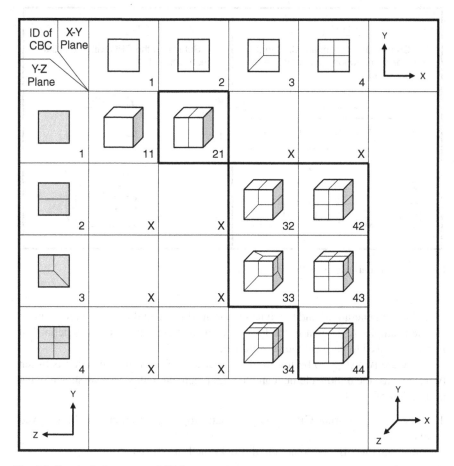

Fig. 4.4 Topological structure of CBCs

This substitution method is named the *CBC substitution approach*, and is presented in the following section. Thereby, the problem of minimising the changes of initial mesh elements can be solved by localising the deformation of the initial meshes only in the area near the boundaries.

4.4 CBC Substitution

As mentioned previously, the mesh elements near the boundary of two adjacent primitives are adjusted locally by either changing the topologies of the elements or moving their nodes along the boundary (for 2D meshing) or within the boundary surface (for 3D meshing) to ensure a correct nodal connectivity. Figure 4.5 illustrates the mechanism of the CBC substitution approach for both the localised mesh alignment

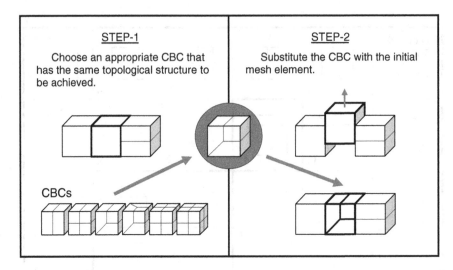

Fig. 4.5 Mesh adjustment by CBC substitution

(initial mesh generation) and the dynamic mesh adjustment through a three-element simple example. As shown in the lower-left of the figure, the six CBCs defined in Fig. 4.4 are used to perform the substitution.

There are two steps to be followed in order to adjust an ill connection between two elements, the middle element and the right element in this case, as shown in the middle-left of Fig. 4.5:

1. Choose an appropriate CBC that has the same topological structure to be achieved and
2. Substitute the CBC with the initial (thick-lined) mesh element.

The modified mesh is shown in the lower-right of the figure. It is worthy to mention that a pre-defined CBC is substituted 'virtually' with a real element to replace its topology. The dimension and orientation of the CBC are adjusted (rotated, scaled or sheared) relatively according to a real situation. More details of the algorithm are presented in the following subsections.

4.4.1 Procedures for CBC Substitution

Assume that two primitives (a fixed primitive F with larger mesh size and a movable primitive M with smaller mesh size) contact each other on a planer surface. The planary surface shared by F and M is called a contacting plane hereafter for mesh adjustment. The basic idea of the procedures is to subdivide the mesh elements of one of the primitives F and M into some smaller ones, if its mesh size is greater than that of the other primitive. Since the process of subdividing F and M is almost the same, only the case where the mesh size of primitive F is greater than that of primitive M is considered here. In other words, the issue now is how to cut the large mesh elements of F, near the boundary, into smaller ones and match them with the elements of M. The procedures consist of five steps.

Step 1 Subdivision of large-size elements
The procedure considered here applies to primitive F only if its mesh size δ^F is greater than two times the mesh size δ^M of primitive M. If it is the case, the mesh elements of F are subdivided to satisfy the following condition: $1 \leq \delta^F/\delta^M \leq 1.5$.

Step 2 Extraction of mesh elements and nodes to be modified
The mesh elements near the contacting surface are evaluated. The nodes that are not shared by both primitives F and M on the contacting surface are detected, and the corresponding elements are extracted based on the geometry data of the primitives. The coordinates of the unmatched nodes must be adjusted, or the extracted elements must be substituted with an appropriate CBC, to achieve one-to-one correspondence between the nodes of F and M.

Step 3 Determination of N-labels
N-labels (node labels) specify the relationships between the nodes of F and M, and provide a base for localising the mesh adjustment. The initial values of the N-labels are attached to the individual nodes of the mesh elements extracted in Step 2. More specifically, an initial value of 1/2 is assigned to the nodes inside of the contacting zone, while keeping 0 as the N-label for other nodes outside of the contacting zone. These initial values are then modified based on the relative positions of the nodes in the primitive F against those in the primitive M (see Sect. 4.4.2 for more details). The value of an N-label $L(V_i)$ describes the following conditions for node V_i of primitive F.

$$L(V_i) = \begin{cases} 0 & \text{Node } V_i \text{ is outside of the zone.} \\ & \text{Not to be adjusted (initial value).} \\ \\ \frac{1}{2} & V_i \text{ is inside of the zone.} \\ & \text{The corresponding node of } M \text{ has not} \\ & \text{been found yet (initial value).} \\ \\ n & V_i \text{ is inside of the zone and has been} \\ & \text{matched with node } V_n \text{ of primitive } M. \\ \\ n + \frac{1}{2} & V_i \text{ is inside of the zone. A new node is} \\ & \text{to be added between } V_n \text{ and } V_{n+1} \text{ of } M. \end{cases} \tag{4.1}$$

Some examples of N-labels are illustrated in Fig. 4.6. After initial values are assigned to the nodes around the contacting border zone, mesh adjustment will focus on the nodes with a label of 1/2. This figure shows a case where node V_i of F has just found its corresponding node V_j of M to be matched.

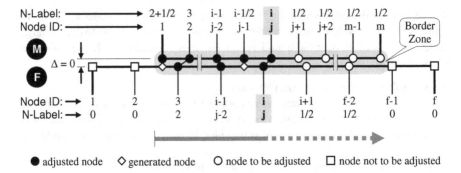

Fig. 4.6 Examples of N-labels during node adjustment

Step 4 Determination of F-labels

F-labels (face labels) attached to the faces of mesh elements specify how the faces should be subdivided in order to adjust the mesh of primitive F against that of primitive M. The F-labels correspond to the numbers in the first row and the first column in Fig. 4.5. Each mesh element of primitive F, therefore, should have two face labels, which specify the ways to subdivide the mesh element in two directions. The value of the F-label S_j of mesh element j is given as follows:

$$
S_j = \begin{cases}
0 & \text{The face is outside of the contacting} \\
  & \text{zone (initial value).} \\
\\
1 & \text{The face should not be further divided} \\
  & \text{(initial value).} \\
\\
2 & \text{The face should be divided into} \\
  & \text{two subfaces.} \\
\\
3 & \text{The face should be divided into} \\
  & \text{three subfaces.} \\
\\
4 & \text{The face should be divided into} \\
  & \text{four subfaces.}
\end{cases}
\tag{4.2}
$$

Each mesh element happening to be adjusted near the contacting border zone should have two F-labels, which are determined based on the N-labels of the mesh element (see Sect. 4.4.3 for more details).

Step 5 Selection of suitable CBCs

A suitable type of CBC is selected for each of the mesh elements that require subdivisions, by referring to the F-labels of the mesh element. The mesh element is then substituted with the selected CBC to replace its topology and to increase or decrease its subdivisions. The nodal coordinates of the

newly modified elements are changed accordingly to match nodes of different
primitives across the zone. This final CBC substitution is applied only to the
mesh elements near the contacting border zone. Most of the initial mesh
elements are not affected by this operation. The coordinates of the nodes in
the new meshes are modified in order to coincide the nodes in both primitives
F and M.

Although it is possible to divide elements into the same size for the two components to achieve an easier nodal matching, the node-splitting problem of moving
components still needs to be solved to allow boundary conditions to pass from one
element to another. As a simplified case, it is covered by the same algorithm.

4.4.2 Determination of N-Labels

The initial values of the N-labels are first given to the nodes of primitives F and M
around their contacting border zone. The initial values are as follows:

$$L^F(V_i^F) = \begin{cases} \frac{1}{2} & \text{if } V_i^F \in V^M \cap V^F \wedge 1 \leq i \leq N(F) \\ 0 & \text{otherwise} \end{cases}$$

$$L^M(V_j^M) = \begin{cases} \frac{1}{2} & \text{if } V_j^M \in V^M \cap V^F \wedge 1 \leq j \leq N(M) \\ 0 & \text{otherwise} \end{cases}$$

(4.3)

The initial values are set to be 1/2 if the nodes are inside of the contacting area
between primitives F and M. Otherwise, they are set to be 0. The nodes with N-labels
of 0s are not considered in the following processes, unless affected by other nodes. For
a node with an N-label of 1/2, it is first adjusted based on the nearest approximation
to the nodes of the adjacent primitive across the contacting border zone. In the case
that no correspondence can be found, a new node is created, as shown in Fig. 4.6, to
ensure one-to-one matched nodal connectivity. The modification of the N-labels is
carried out from the node with the smallest ID number. Figure 4.7 summarises the
procedures of how to find or create a node in seven possible cases.

In the figure, the white nodes and the black nodes show the nodes outside and
inside of the contacting area, respectively. The grey nodes are the nodes to be moved
for local adjustment, and the diamond nodes to be created. For example, the node
$V_j^M \notin V^M \cap V^F$ is moved to and matched with node $V_i^F \in V^M \cap V^F$ in the
primitive F in case 1. Their N-labels are changed to i and j, respectively, to reflect
this adjustment. On the other hand, as shown in the case of 5(a) of Fig. 4.7, the node
$V_{j+1}^M \in V^M \cap V^F$ at the right edge of primitive M is left unmatched, and a new
node $V_i'^F$ is therefore created between nodes V_i^F and V_{i+1}^F to pair with V_{j+1}^M, instead
of moving node V_{i+1}^F backward. This simplifies the algorithm processing effectively
for unidirectional node adjustment (from left to right along the border line).

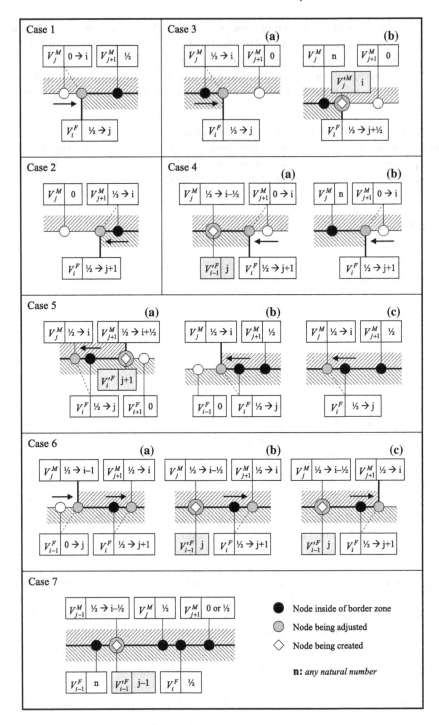

Fig. 4.7 Seven possible cases of node adjustment

The algorithm of N-label assignment is developed to cover those possible cases illustrated in Fig. 4.7. It is worthy to mention that this operation is localised in the contacting border zone and performed from one end to the other along the border. Its processing procedure is given in the following C-like language:

for $\forall V_i^F (V_i^F \in V^F \wedge L^F(V_i^F) == \frac{1}{2})$ do

    search a $V_j^M \in V^M$ that **Is_Between**$(V_j^M, V_{j+1}^M) \ni V_i^F$;

    sw $\Longleftarrow$ **Check_Status**(V_i^F);

    switch$(sw) \rightsquigarrow$ (See Fig. 4.7)

        case 1: $L^M(V_j^M) == 0 \wedge$ **Is_Near**$(V_j^M) \ni V_i^F$

            Move_To(V_j^M, V_i^F);

            $L^M(V_j^M) = i$;

            $L^F(V_i^F) = j$;

            break;

        case 2: $L^M(V_j^M) == 0 \wedge$ **Is_Near**$(V_{j+1}^M) \ni V_i^F$

            Move_To(V_{j+1}^M, V_i^F);

            $L^M(V_{j+1}^M) = i$;

            $L^F(V_i^F) = j + 1$;

            break;

        case 3: $L^M(V_{j+1}^M) == 0 \wedge$ **Is_Near**$(V_j^M) \ni V_i^F$

            if $L^M(V_j^M) == \frac{1}{2}$

                Move_To(V_j^M, V_i^F);

                $L^M(V_j^M) = i$;

                $L^F(V_i^F) = j$;

            else

                Create_Between$(V_j'^M, V_j^M, V_{j+1}^M)$;

                $L^M(V_j'^M) = i$;

                $L^F(V_i^F) = j + L^F(V_i^F)$;

            endif

            break;

        case 4: $L^M(V_{j+1}^M) == 0 \wedge$ **Is_Near**$(V_{j+1}^M) \ni V_i^F$

            if $L^M(V_j^M) == \frac{1}{2}$

                Create_Between$(V_{i-1}'^F, V_{i-1}^F, V_i^F)$;

$$L^F(V_{i-1}'^F) = j;$$
$$L^M(V_j^M) = i - L^M(V_j^M);$$

endif

Move_To(V_{j+1}^M, V_i^F);

$$L^M(V_{j+1}^M) = i;$$
$$L^F(V_i^F) = j + 1;$$

break;

case 5: $L^M(V_j^M) == \frac{1}{2} \wedge$ **Is_Near** $(V_j^M) \ni V_i^F$

if $L^F(V_{i+1}^F) == 0$

Create_Between$(V_i'^F, V_i^F, V_{i+1}^F)$;

$$L^F(V_i'^F) = j + 1;$$
$$L^M(V_{j+1}^M) = i + L^M(V_{j+1}^M);$$

endif

Move_To(V_i^F, V_j^M);

$$L^F(V_i^F) = j;$$
$$L^M(V_j^M) = i;$$

break;

case 6: $L^M(V_j^M) == \frac{1}{2} \wedge$ **Is_Near**$(V_{j+1}^M) \ni V_i^F$

Move_To(V_i^F, V_{j+1}^M);

$$L^F(V_i^F) = j + 1;$$
$$L^M(V_{j+1}^M) = i;$$

if $L^F(V_{i-1}^F) == 0 \wedge$ **Is_Near**$(V_j^M) \ni V_{i-1}^F$

Move_To(V_{i-1}^F, V_j^M);

$$L^F(V_{i-1}^F) = j;$$
$$L^M(V_j^M) = i - 1;$$

else

Create_Between$(V_{i-1}'^F, V_{i-1}^F, V_i^F)$;

$$L^F(V_{i-1}'^F) = j;$$
$$L^M(V_j^M) = i - L^M(V_j^M);$$

endif

break;

endswitch

if $\mathbf{Exist}(V_{j-1}^M) \wedge L^M(V_{j-1}^M) == \frac{1}{2}$ (case 7 in general)

$\quad\mathbf{Create\_Between}(V_{i-1}'^F, V_{i-1}^F, V_i^F);$

$\quad L^F(V_{i-1}'^F) = j - 1;$

$\quad L^M(V_{j-1}^M) = i - L^F(V_i^F);$

endif

endfor

where **Is_Between, Check_Status, Is_Near, Move_To, Create_Between** and **Exist** are defined macro functions for nodal status checking and node alignment.

4.4.3 Determination of F-Label

The values of F-labels of a primitive are largely determined based on the N-labels of another primitive across the border. In other words, the F-labels S^F are determined directly by the N-labels L^M, and S^M by L^F vice versa. The F-labels should be considered in two directions; they are, the direction designated by P which is the direction of the relative motion between the primitives F and M, and the direction designated by V which is perpendicular to P and the normal vector of the contacting surface shared by F and M. Figure 4.8 shows a typical example of the decision process of F-labels determination.

The first (top) row in the figure gives a schematic illustration of the N-labels of primitive M and the nodal connectivity between the nodes of M and F across the

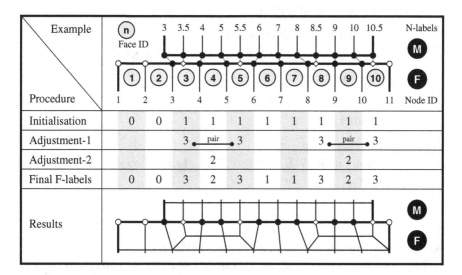

Fig. 4.8 F-label determination through an example

border. The initial values of the F-labels are assigned to each mesh element of F as shown in the second row (Initialisation) in the figure, using the following equations. Since the mesh size of F is larger than that of M, only the mesh elements of F should be subdivided in this case. The F-labels of the primitive F are therefore considered here as an example.

$$S^F(i) = \begin{cases} 0 & \text{if } N\text{-label of } i\text{th node } L^F(V_i^F) = 0 \\ 1 & \text{otherwise} \end{cases}$$

$$S^M(j) = \begin{cases} 0 & \text{if } N\text{-label of } j\text{th node } L^M(V_j^M) = 0 \\ 1 & \text{otherwise} \end{cases}$$

(4.4)

The N-labels of M are then used to adjust the initial F-labels of F. As shown in the third row (Adjustment-1), some F-labels are changed from 1 to 3 by following the procedure outlined below:

for $\forall V_j^M \in V^M$ do

    if $L^M(V_j^M) == \xi + \frac{1}{2} \wedge \xi \in N^F$

        $S^F(\xi) = 3$;

    endif

endfor

where ξ is a parameter giving the ID numbers of the nodes, and N^F is the node set of F. The elements with F-labels of 3 are paired together (see Fig. 4.8). The F-labels of those elements in between the pairs are changed from 1 to 2 as shown in the fourth row (Adjustment-2) of the figure, by applying the following procedure:

for $\forall i \in N^F$ that $S^F(i) \geq 1$ do

    search an i that $S^F(i) == 3$;

    search a λ that $S^F(i + \lambda) == 3 \wedge (i + \lambda) \in N^F$;

    for $\forall \gamma (1 < \gamma < \lambda)$ do

        $S^F(i + \gamma) = 2$;

    endfor

endfor

search an i if that $S^F(i) == 0 \wedge S^F(i - 1) == 2$ do

    $S^F(i) = 3$;

endall

The final results are summarised in the last row in Fig. 4.8.

    The procedures mentioned above are carried out along the two directions designated by P and V of primitive F in order to obtain the F-labels S_P^F and S_V^F for three-dimensional mesh adjustment. The face labels S_P^M and S_V^M of primitive M can be obtained by applying the same procedures, based on the N-labels of F.

The aforementioned procedures make the F-labels determination easy and effective, without introducing any forbidden or ill cases.

4.4.4 Selection of Suitable CBCs

Suitable CBCs are selected for those mesh elements whose nodes disagree with each other, by referring to the F-labels of the elements in the directions of P and V. Assuming that S_P and S_V are two sets of F-labels of the elements in the two directions, the ID number of an appropriate CBC can be obtained by using the following equations:

$$ID_{CBC} = S_P \times S_V \tag{4.5}$$

or

$$ID_{CBC} = \{(S_P(x), S_V(y)) | S_P(x) \in S_P \wedge S_V(y) \in S_V\} \tag{4.6}$$

The individual mesh elements, which are identified for subdivisions, are then simply substituted with the CBCs to modify their topologies, and the nodal coordinates of the replaced elements are adjusted properly to achieve a smooth nodal connectivity.

For example, assuming that the F-label $S_P^F(x) = 3$ for the xth element of F in the direction of P and $S_V^F(y) = 2$ for the yth element of F in the direction of V, the ID of an appropriate CBC to be substituted with the (x, y)th initial element can be found as follows:

$$ID_{CBC}^F(x, y) = (S_P^F(x), S_V^F(y)) = (3, 2) \overset{\text{def}}{\Longrightarrow} 32.$$

It is worth mentioning that the elements with ID of 11 are not changed in topology. This can largely eliminate the number of elements being affected during dynamic mesh adjustment and hence increase the efficiency.

4.5 Extension of CBC Substitution

4.5.1 Concept of Two-Space Meshing

The CBC substitution approach described in the previous section is basically developed for the geometry with plane surfaces. The algorithm is proven to be simple and effective when primitives contact with each other on planary surfaces. An extension of this approach is needed to handle the more complex cases of mesh adjustment for arbitrary primitives with curved surfaces, or curved primitives.

Mapping technique in projective geometry [83] is adopted for integrating the curved primitives with the CBC substitution approach. Figure 4.9 illustrates the concept of two-space meshing and the extended CBC substitution through a meshing example of metal-milling.

As shown in Fig. 4.9, the geometric model of a tool-workpiece is first generated in real object space R^3, where all the actual models are represented. Simultaneously, the geometric model is transformed to its corresponding projective space RP^3, where initial mesh elements are generated to each of the projected primitives (projective images). Since the curved surfaces can be transformed to planary ones by using appropriate mapping functions, the needed mesh adjustment can be done easily in RP^3 by the proposed CBC substitution approach. Upon completion, the projected primitives are transformed inversely to R^3, by selecting the corresponding inverse mapping functions properly. It should be emphasised that the mesh adjustment is done in RP^3 while the FEM analysis or simulation is performed in R^3, by repeating the process of Step-3, as shown in Fig. 4.9, whenever relative motion occurred between the primitives.

4.5.2 Mapping and Inverse Mapping

Mapping and inverse mapping bridge the real object space and its projective space, making the two-way projective transformations between a real model and its image possible. Figure 4.10 shows the relationship between mapping and inverse mapping, and how the mapping functions can be found. A circle $C^2 = \{(x, y) \in R^2 | x^2 + y^2 \le 1\}$ in two-dimensional Euclidean space R^2 (real object space) and a square $S^2 = \{(\xi, \eta) \in RP^2 | |\xi| \le 1, |\eta| \le 1\}$ in projective space RP^2 are taken as an example.

Since both circle and square are compact sets, continuous projection exists between C^2 and S^2. The procedures of finding a mapping function $f : C^2 \to S^2$ and its inverse mapping function $g : S^2 \to C^2$ are as follows:

1. Draw a projection line outward from any point P that is inside of both C^2 and S^2.
2. Find the intersections S and I with C^2 and S^2, respectively.
3. Derive the relationship between S and I so as to establish the functions f and g.

It is not difficult to derive the mapping and inverse mapping functions for this simple example as:

$$f(x, y) = \begin{cases} (\lambda x, \lambda y) \; \lambda = \frac{\sqrt{x^2+y^2}}{\max\{|x|,|y|\}} & (x, y) \ne (0, 0) \\ (0, 0) & (x, y) = (0, 0) \end{cases}$$

$$g(\xi, \eta) = \begin{cases} (\lambda \xi, \lambda \eta) \; \lambda = \frac{\max\{|\xi|,|\eta|\}}{\sqrt{\xi^2+\eta^2}} & (\xi, \eta) \ne (0, 0) \\ (0, 0) & (\xi, \eta) = (0, 0) \end{cases}$$

(4.7)

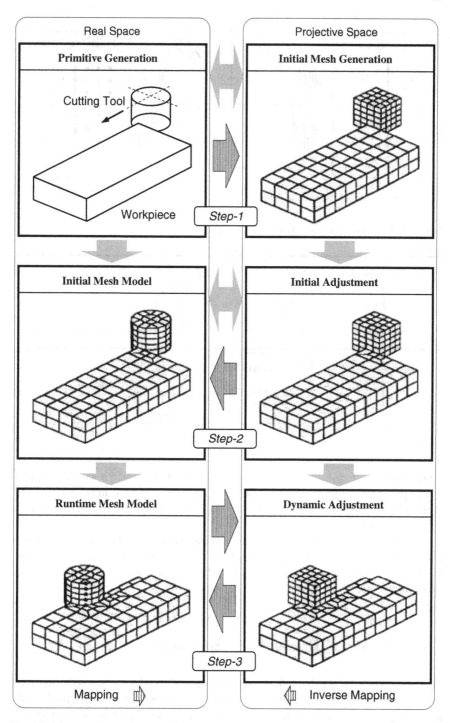

Fig. 4.9 Mapping for extended CBC substitution

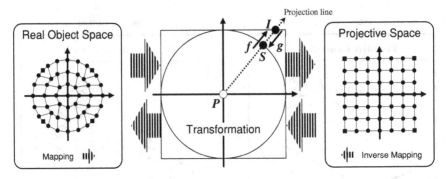

Fig. 4.10 Mapping and inverse mapping

Table 4.1 Typical mapping and inverse mapping functions

Cylinder

$$f(x, y, z) = \begin{cases} (\lambda x, \lambda y, z), & \lambda = \frac{\sqrt{x^2+y^2}}{\max\{|x|,|y|\}} & (x, y, z) \neq (0, 0, z) \\ (0, 0, z) & (x, y, z) = (0, 0, z) \end{cases}$$

$$g(\xi, \eta, \zeta) = \begin{cases} (\lambda \xi, \lambda \eta, \zeta), & \lambda = \frac{\max\{|\xi|,|\eta|\}}{\sqrt{\xi^2+\eta^2}} & (\xi, \eta, \zeta) \neq (0, 0, \zeta) \\ (0, 0, \zeta) & (\xi, \eta, \zeta) = (0, 0, \zeta) \end{cases}$$

Sphere

$$f(x, y, z) = \begin{cases} (\lambda x, \lambda y, z), & \lambda = \frac{\sqrt{x^2+y^2+z^2}}{\max\{\|x\|,\|y\|,\|z\|\}} & (x, y, z) \neq (0, 0, 0) \\ (0, 0, 0) & (x, y, z) = (0, 0, 0) \end{cases}$$

$$g(\xi, \eta, \zeta) = \begin{cases} (\lambda \xi, \lambda \eta, \zeta), & \lambda = \frac{\max\{\|\xi\|,\|\eta\|,\|\zeta\|\}}{\sqrt{\xi^2+\eta^2}} & (\xi, \eta, \zeta) \neq (0, 0, 0) \\ (0, 0, 0) & (\xi, \eta, \zeta) = (0, 0, 0) \end{cases}$$

Since $f \cdot g = g \cdot f = 1$, if regarding (ξ, η) as (x, y), the circle C^2 is said topological isomorphic to the square S^2 or $C^2 \cong S^2$. This fact is of vital importance to the uniformity of projective transformation between R^2 and RP^2 without any ill cases (e.g. $f : X \rightarrow Y, g : Y \rightarrow Z$, but $X \neq Z$).

Two pairs of commonly used mapping and inverse mapping functions are defined and listed in Table 4.1. The mapping functions of other geometries (cube, cone, torus or any other analytical surfaces) can be found in the same way. These functions are used to calculate the nodal coordinates during mappings.

4.5.3 Algorithm of Extended CBC Substitution

As shown in Fig. 4.9, the dynamic mesh adjustment of a model with curved primitives includes projective transformations (mapping and inverse mapping), initial mesh alignment and dynamic mesh adjustment. The processing algorithm of the extended approach is divided into three steps accordingly and described in C-like language in detail.

Step 1 Establishing a projective space and mapping the original model to the projective space.

The procedure in Step 1 is to transform all curved primitives from R^3 to RP^3 through well-defined mapping functions. Initial meshes are generated to the projected primitives (images) before meshes are aligned across borders in the next step.

for $\forall pr$ $(pr \in R^3 \wedge pr \in \{\textbf{Curved\_Primitive\_Set}\})$ do

  $M_{id} \Longleftarrow \textbf{Select\_Mapping\_Function}\ (pr)$;

  $\textbf{Map\_To}\ (M_{id}, pr, I_{pr})$;

  $\textbf{Generate\_Initial\_Mesh}\ (I_{pr})$;

  $\textbf{Push}\ I_{pr} \Longleftarrow RP^3$;

endfor

for $\forall pr$ $(pr \in R^3 \wedge pr \notin \{\textbf{Curved\_Primitive\_Set}\})$ do

  $\textbf{Generate\_Initial\_Mesh}\ (pr)$;

  $\textbf{Duplicate}\ pr \Longleftarrow RP^3$;

endfor

Step 2 Adjusting initial meshes and mapping them inversely to the real space.

The initial meshes generated in the previous step are adjusted locally in the projective space. A complete mesh model for analysis is obtained by inverse mapping. Any meshing data changes should be reflected in both spaces.

for $\forall I_{pr}$ $(I_{pr} \in RP^3)$ do

  $\textbf{Adjust\_Mesh}\ (I_{pr}, CBC)$;

  $\textbf{Refresh}\ I_{pr} \rightarrow RP^3$;

  $M_{id} \Longleftarrow \textbf{Select\_Inverse\_Mapping\_Function}\ (I_{pr})$;

  $\textbf{Inverse\_Map\_To}\ (M_{id}, I_{pr}, pr)$;

  $\textbf{Refresh}\ pr \rightarrow R^3$;

endfor

Step 3 Dynamic mesh adjustment between the two spaces.

The relative positions among model components may vary with time during dynamic FEA or simulation. It is necessary to modify the FEM mesh model dynamically whenever the relative positions are changed. The dynamic mesh adjustment is described as follows:

whenever $(\textbf{Relative\_Motion\_Occured} == true)$

for $\forall pr$ $(pr \in R^3)$ do

status $\Longleftarrow$ **Check_Motion_Status** (pr);

if (status == **Moved**)

$I_{pr} \Longleftarrow$ **Select_Projected_Object** (pr);

Adjust_Mesh (I_{pr}, CBC);

Refresh $I_{pr} \rightarrow RP^3$;

$M_{id} \Longleftarrow$ **Select_Inverse_Mapping_Function** (I_{pr});

Inverse_Map_To (M_{id}, I_{pr}, pr);

Refresh $pr \rightarrow R^3$;

endfor

Special attentions to those nodes on the border zones of the primitives should be given to the inverse mapping in Steps 2 and 3. Figure 4.11 illustrates the nodal connectivity before and after the inverse mapping. Since the nodes indicated by white circles differ in nodal connectivity after the inverse mapping, each of them are divided into two and assigned to the primitives that previously shared the node. The other nodes given by ⊙ remain unchanged.

Fig. 4.11 Node division during inverse mapping

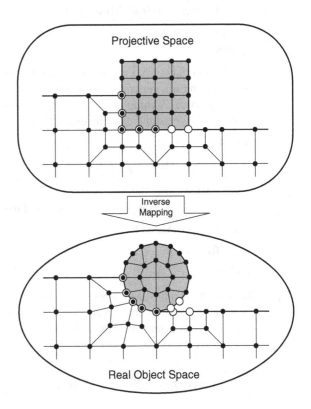

> Whether to divide a node or not is simple—it is judged by the nodal connectivity of the sharing primitives in the real object space R^3.

After inverse mapping as shown in Fig. 4.11, there are cases that three nodes of an element share one arc segment. If the three nodes approach to collinear, the accuracy of FEA may decrease. To prevent it from happening, we use 20-node elements instead of 8-node elements for FEA calculations.

4.6 Application of CBC Substitution to Machine Tool Models

4.6.1 Considerations of a New Data Structure

The procedures of node adjustment and CBC substitution mentioned above should be carried out whenever relative motions between machine components take place. It is a very time-consuming process to change the finite element model of a machine tool when it is put under running state, because of the frequent occurrence of the relative motions and the great quantities of computation for nodal coordinates. It will become even worse when a large number of mesh elements or very complex geometries of machine tools are encountered. In fact, only the nodal coordinates of those moved components are needed to be recalculated when some kinds of motions occurred. Therefore, it is urgently required to establish a suitable data structure to represent the geometries, the relative motions and other necessary information about the machine tools for realising an efficient and effective dynamic FEM mesh generation.

In recognition of this problem, a new data structure is proposed in this section based on the data structure proposed in Chap. 3, aiming at minimising the computation and speeding up the substitution process. A tree structure is adopted for the basic logical organisation of the proposed data structure, and two types of relational operations are used to extract the necessary information for the CBC substitution. Further details about the representation of the machine tool models based on the data structure are given in the following subsections.

4.6.2 Representation of Machine Tool Models

The CBC substitution approach explained so far refers to the information about the geometries and the relative motions of the machine tools, such as the shapes and the dimensions of the primitives, the connecting relations among the primitives, the positions and the motions of the primitives. Once the solid model data of a machine tool is created, its FEM model of the initial state can be obtained automatically from the solid model data, by applying the CBC substitution approach. The CBC substitution is also applied to change the FEM meshes of the machine tools of the

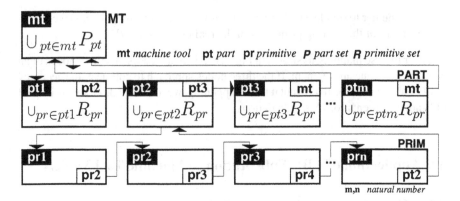

Fig. 4.12 Hierarchical data structure of machine tools

initial state whenever positions of any components are changed due to, for example, feeding motions and cutting motions.

The machine tool model considered here consists of three hierarchical elements based on the tree structure; they are, machine tools, parts and primitives. The parts are treated as the rigid elements from the viewpoint of the relative motions of the machine tools. The tree structure adopted here describes the overall logical organisation of data, which gives the names of the data entities and attributes, and specifies the associations that exist between them. A tree is made up of a multilevel group of elements called *nodes*. A node is nothing more than a point at which subsidiary data originate. Figure 4.12 gives a schematic illustration of the data structure of the machine tools consisting of the three types of components. The pointers are adopted to represent the one-to-many relationships between the machine tools and the parts, and those between the parts and the primitives.

In the figure, MT, PART and PRIM represent the three hierarchical elements, or the names of the data entities: Machine Tools, PARTs and PRIMitives, respectively. Data presented in a tree structure must meet the following two conditions; they are

1. The tree must have a single *root* node and
2. All nodes other than the root node must be related to one and only one higher-level node.

Here, MT serves as the root node, and PART and PRIM the leaf nodes of the tree. It should be noticed that the *node* mentioned here is a quite different concept from the *node* of FEM mesh elements, although they share the same name.

The contents of the individual entities are summarised in Fig. 4.13. The contacting information in PRIM describes the contacting relationships between the primitives. The information about the relative motions between the parts is given in the information about joint in PART.

Fig. 4.13 Attributes of entities

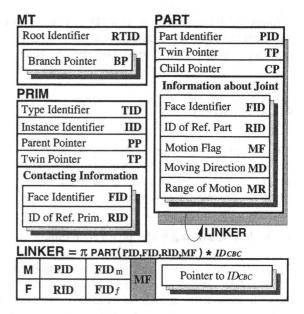

4.6.3 Utilisation of Machine Tool Model in Meshing

The information about the nodes and the mesh elements of the primitives is retrieved and referred to in the CBC substitution approach according to the data structures shown in Figs. 4.12 and 4.13. For instance, the information about the relative motions and that about contacting surfaces between a pair of primitives is stored separately in the entities PART and PRIM. The 'Information about Joint' and 'Contacting Information' appear as a *repeating group*, that is, they occur more than once in PART and PRIM, respectively. The data describing the relative motions and contacting surfaces should be gathered and put together in one entity for the ease of the CBC substitution. An entity (or relation) LINKER is therefore generated from the entity PART and the set ID_{CBC} as follows:

$$\text{LINKER} = \pi \text{ PART}(\textbf{PID, FID, RID, MF}) * ID_{CBC} \qquad (4.8)$$

where π and $*$ correspond to the *project* operator and the *join* operator in the DBMS (*Data Base Management System*) [84], respectively. **PID, FID, RID** and **MF** within parentheses are the names of the data items which are needed to be extracted from PART. This entity gives the information required in the CBC substitution approach.

Suppose a case where a relative motion takes place on the surface specified by **FID** between the components specified by **PID** and **RID**. The meshes and the nodes of the components are first modified and adjusted to obtain a set of ID_{CBC} by applying the CBC substitution approach. The value of corresponding item **MF** in the newly created entity LINKER is secondly set to be 1. The coordinates of the nodes on the surfaces which satisfy the condition **MF** $= 1$ are finally calculated, and the values of **MF** are set to be 0 after the calculation. The procedures mentioned above are carried out whenever the relative positions between the components are changed.

4.7 Case Studies

The concept and algorithms for solid modelling and finite elements generation and adjustment have been implemented into software tools. These tools are applied to several case studies for automatic and dynamic meshing of the machine tools, among which two are introduced in this section.

Example 1: A Moving Slider on a Fixed Table

Figure 4.14 shows one example of modelling and dynamic FEM mesh generation in which a slider moves on top of a fixed table. It is first recognised that the meshes of the table are to be subdivided as its mesh size is larger than that of the slider. The node adjustment of meshes along Y-axis is first carried out in the middle layer in X direction in X-Y plane. The node adjustment of meshes along Z-axis in the X-Z plane is conducted in the top layer of the table. Figure 4.14a–h illustrates eight resultant meshes generated dynamically and automatically, when the slider moves along the Z-axis on the upper surface of the table.

Example 2: A Vertical Machining Centre

Figure 4.15 shows another example of modelling and dynamic FEM mesh generation of a vertical machining centre with relative motions. It is first recognised that the relative motions take place between the spindle head and the column, and between the table and the base. Since the mesh sizes of the fixed parts (the column and the base) are greater than those of the movable parts (the spindle head and the table, respectively), the meshes of the fixed parts are, therefore, subdivided and adjusted in order to realise the one-to-one correspondence of the nodes between the fixed parts and the movable parts. Figure 4.15a shows the initial FEM model of the machining centre obtained from its solid model, and b illustrates one of the generated FEM meshes adjusted dynamically between the fixed parts and the movable parts after the relative motions occurred.

The results of the case studies given in Figs. 4.14 and 4.15 demonstrate that suitable finite element models can be generated automatically and dynamically when the relative motions take place between different components by using the CBC substitution approach.

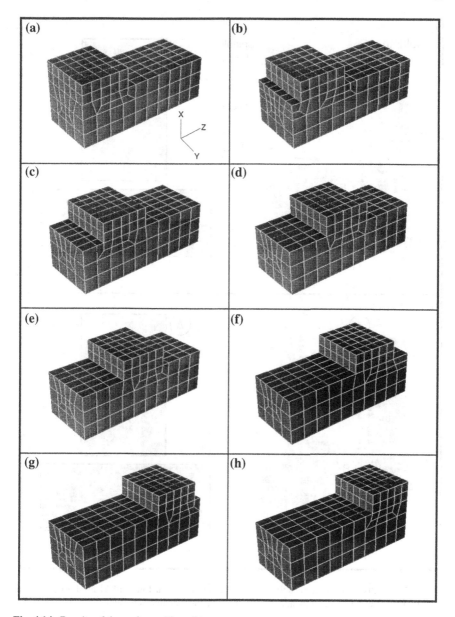

Fig. 4.14 Results of dynamic meshing of two components

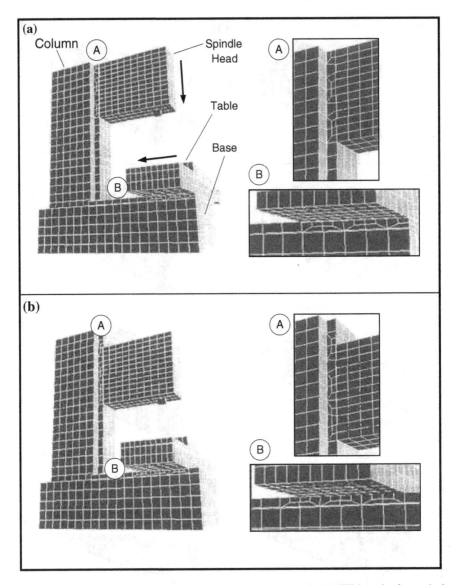

Fig. 4.15 Results of dynamic meshing of a machining centre. **a** Initial FEM mesh of a vertical machining centre. **b** Dynamically adjusted mesh while the machine is moving

4.8 Concluding Remarks

In this chapter, existing finite element mesh generation methods in the literature are reviewed and classified. Seven classes of mesh generation approaches are identified. Only four of them are found to be automatic. However, there is still no fully automated mesh generation methods existed for 3D models in motion, especially for machine tools under running state. A new method called CBC substitution approach is proposed and introduced to carry out the automatic and dynamic 3D mesh generation for the FEM analysis of the machine tools, based on their solid model data. A new data structure is also introduced for the CBC substitution approach aimed at minimising the computational effort and speeding up the substitution process. The models of the machine tools are represented by the hierarchical data structure consisting of three entities: the machine tools, the parts and the primitives.

The FEM meshes of hexahedral elements are initially generated for individual primitives constituting the parts and the machine tools. The meshes of the individual primitives are changed locally at the contacting planes shared by a pair of primitives in order to make a one-to-one correspondence between the nodes of the FEM meshes of the primitives. The CBC substitution approach is applied to change the FEM meshes of the primitives at the contacting planes, especially when relative motions occurred. The method proposed is effective to generate the FEM meshes of the machine tools automatically and dynamically in the operating conditions where machine components move relatively.

The following remarks are concluded through the meshing case studies:

1. The FEM meshes can be generated and modified automatically and dynamically at the contacting planes between the primitives by applying the CBC substitution approach.
2. During the process of mesh adjustment, the necessary information is set and stored in N-labels and F-labels, and processed automatically without any user intervention.
3. The FEM meshes and the nodes of the meshes are changed and adjusted locally near the contacting planes between the primitives, and therefore, only one or two layers of the FEM meshes in the primitives are changed by the CBC substitution.
4. The information about the geometries and the relative motions of machine tools can be obtained directly from the hierarchical data structure of the machine tools. The FEM meshes of the machine tools are generated automatically, dynamically and effectively based on the data structure when the relative motions take place.
5. The CBC substitution approach is simple, logical and straightforward. By using this method, a constructive FEM model with reasonably well-shaped hexahedral elements can be created by simply assembling the primitives.

The CBC substitution approach can be extended to cover meshing requirements of objects with curved surfaces, by using mapping and inverse mapping techniques. The meshing process is performed after projecting the curved objects to a projective space where the curved surfaces are represented by planary ones. Upon completion of the mesh adjustment, they are inversely mapped back to the real object space for finite element analysis. The mapping and meshing operations are carried out iteratively whenever a relative motion happens during FEM calculations.

References

1. W.K. Liu, S. Jun, S. Li, J. Adee, T. Belytschko, Reproducing Kernel particle methods for structural dynamics. Int. J. Numer. Meth. Eng. **28**, 1655–1679 (1995)
2. Y.Y. Lu, T. Belytschko, L. Gu, A new implementation of the element free Galerkin method. Comput. Methods Appl. Mech. Eng. **113**, 397–414 (1994)
3. E. Oñate, S. Idelsohn, A mesh-free finite point method for advective-diffusive transport and fluid flow problems. Comput. Mech. **21**, 283–292 (1998)
4. T. Liszka, J. Orikisz, Finite difference method at arbitrary irregular grids and its application in applied mechanics. Comput. Struct. **11**, 83–95 (1980)
5. Y.X. Mukherjee, S. Mukherjee, The boundary node method for potential problems. Int. J. Numer. Meth. Eng. **40**, 797–815 (1997)
6. C.A. Duarte, J.T. Oden, An h-p adaptive method using clouds. Comput. Methods Appl. Mech. Eng. **139**, 237–262 (1996)
7. T. Zhu, J. Zhang, S.N. Atluri, A meshless local boundary integral equation (LBIE) method for solving nonlinear problems. Comput. Methanics **22**, 174–186 (1998)
8. N.R. Aluru, G. Li, Finite cloud method: A true meshless technique based on a fixed reproducing Kernel approximation. Int. J. Numer. Meth. Eng. **50**, 2373–2410 (2001)
9. J.U. Turner, Accurate solid modeling using polyhedral approximations. IEEE Comput. Graph. Appl., pp. 14–27 (1988)
10. W.H. Chen, J.T. Yeh, Finite element analysis of finite deformation contact problems with friction. Comput. Struct. **29**(3), 423–436 (1988)
11. J.M. Guedes, N. Kikuchi, Preprocessing and postprocessing for materials based on the homogenization method with adaptive finite element methods. Comput. Methods Appl. Mech. Eng. **83**, 143–198 (1990)
12. L. Wang, T. Moriwaki, An approach to dynamic finite element mesh generation for machines with relative motions. Mem. Grad. School Sci. Technol., Kobe Univ. **11-A** (1993)
13. A. Denayer, Automatic generation of finite element meshes. Comput. Strcutures **9**, 359–364 (1978)
14. L.R. Herrmann, Laplacian-isoparametric grid generation scheme. J. Eng. Mech. Div. Proc. Am. Soc. Civil Eng. **102**(EM5):10 (1976)
15. Z.J. Cendes, D. Shenton, H. Shahnasser, Magnetic field computation using delaunay triangulation and complementary finite element methods. IEEE Trans. Mag. **MAG–19**, 6 (1983)
16. T.I. Boubez, W.R.J. Funnell, D.A. Lowther, A.R. Pinchuk, P.P. Silvester, Mesh generation for computational analysis. J. Comput. Aided Eng. **10**, 190–201 (1986)
17. W. Brostow, J.P. Dussault, Construction of voronoi polyhedra. J. Comput. Phys. **29**, 81–92 (1978)
18. P.J. Green, R. Sibson, Computing dirichlet tessellations in the plane. Comput. J. **21**(2), 168–173 (1977)
19. B. Wordenweber, Volume triangulation. in *Technical Report-CAD Group Document (University of Cambridge)*, No. 110, (1980)
20. B. Wordenweber, Finite element mesh generation. Comput. Aided Des. **16**(5), 285–291 (1984)
21. B.G. Baumgart, Geometric modeling for computer vision. in *Report No. CS-463 Stanford Artificial Intelligence Laboratory, Computer Science Department*, Stanford, USA, 1974.
22. J. Suhara, J. Fukuda, *Automatic Mesh Generation for Finite Element Analysis. Advances in Computational Methods in Structural Mechanics and Design* (UAH Press, Huntsville, Alabama, USA, 1972)
23. A.O. Moscardini, B.A. Lewis, M. Cross, AGTHOM-automatic generation of triangular and higher order meshes. Int. J. Numer. Meth. Eng. **19**, 1331–1353 (1983)
24. R.D. Shaw, R.G. Pitchen, Modifications to the SUHARA-FUKUDA method of network generation. Int. J. Numer. Meth. Eng. **12**, 93–99 (1978)
25. S.H. Lo, A new mesh generation scheme for arbitrary planar domains. Int. J. Numer. Meth. Eng. **21**, 1403–1426 (1985)

26. B.A. Lewis, J.S. Robinson, Triangulation of planar rigions with applications. Comput. J. **21**(4), 324–332 (1977)

27. C.O. Frederick, Y.C. Wong, F.W. Edge, Two-dimensional automatic mesh generation for structural analysis. Int. J. Numer. Meth. Eng. **2**, 133–144 (1970)

28. J.M. Nelson, A triangulation algorithm for arbitrary planar domains. Appl. Math. Modeling **2**, 151–159 (1978)

29. M.B. McGirr, D. Corderoy, P. Easterbrook, A. Hellier, A new approach to automatic mesh generation in the continuum. in *Proceedings of 4th International Conference Australia Finite Element Method*, Melbourne, Australia, pp. 36–40, (1982)

30. J.C. Cavendish, Automatic triangulation of arbitrary planar domains for the finite element method. Int. J. Numer. Meth. Eng. **8**, 679–696 (1974)

31. E.A. Sadek, A scheme for the automatic generation of triangular finite elements. Int. J. Numer. Meth. Eng. **15**, 1813–1822 (1980)

32. Y. Liu, K. Chen, A two-dimensional mesh generator for variable order triangular and rectangular elements. Comput. Struct. **29**(6), 1033–1053 (1988)

33. N. Van Phai, Automatic mesh generation with tetrahedron elements. Int. J. Numer. Meth. Eng. **18**, 273–289 (1982)

34. J.C. Cavendish, D.A. Field, W.H. Frey, An approach to automatic three-dimensional finite element mesh generation. Int. J. Numer. Meth. Eng. **21**, 329–347 (1985)

35. N.A. Calvo, S.R. Idelsohn, All-hexahedral element meshing: Generation of the dual mesh by recurrent subdivision. Comput. Methods Appl. Mech. Eng. **182**, 371–378 (2000)

36. S.H. Lo, Finite element mesh generation over curved surfaces. Comput. Struct. **29**(5), 731–742 (1988)

37. F. Cheng, J.W. Jaromczyk, J.R. Lin, S.S. Chang, J.Y. Lu, A parallel mesh generation algorithm based on the vertex label assignment scheme. Int. J. Numer. Meth. Eng. **28**, 1429–1448 (1989)

38. B.K. Karamete, M.W. Beall, M.S. Shephard, Triangulation of arbitrary polyhedra to support automatic mesh generator. Int. J. Numer. Meth. Eng. **49**, 167–191 (2000)

39. S. Dey, R.M. O'Bara, M.S. Shephard, Towards curvilinear meshing in 3D: The case of quadratic simplices. Comput. Aided Des. **33**, 199–209 (2001)

40. A. Kela, R. Perucchio, H.B. Voelcker, Toward automatic finite element analysis. Comput. Mech. Eng. **5**, 1 (1986)

41. W.C. Thacker, A. Gonzalez, G.E. Putland, A method for automating the construction of irregular computational grids for storm surge forecast models. J. Comput. Phys. **37**, 371–387 (1980)

42. N. Kikuchi, Adaptive grid-design methods for finite element analysis. Comput. Methods Appl. Mech. Eng. **55**, 129–160 (1986)

43. E.A. Heighway, C.S. Biddlecombe, Two-dimensional automatic triangular mesh generation for the finite element electromagnetics package PE2D. IEEE Trans. on Mag. **MAG–18**(2), 594–598 (1982)

44. K.K. Wang, N. Hashimoto, Test and evaluation of TIPS-1 system. in *Technical Report MME-04* (Cornell University, Ithaca, NY, USA, 1981)

45. T. Akiyama, K.K. Wang, A TIPS-1 based CAD program for mold design. in *Proceedings of 9th North American Manufacturing Research Conference* (1981)

46. M.A. Yerry, M.S. Shephard, A modified quadtree approach to finite element mesh generation. IEEE Comput. Graph. Appl., pp. 39–46, (1983)

47. H. Samet, The quadtree and related hierarchical data structures. ACM Comput. Surv. **16**(2), 187–260 (1984)

48. S.F. Yeung, M.B. Hsu, A mesh generation method based on set theory. Comput. Strcuct. **3**, 1063–1077 (1973)

49. R. Haber, M.S. Shephard, J.F. Abel, R.H. Gallagher, D.P. Greenberg, A general two-dimensional graphical finite element preprocessor utilizing discrete transfinite mappings. Int. J. Numer. Meth. Eng. **17**, 1015–1044 (1981)

50. C.A. Hall, *Transfinite Interpolation and Applications to Engineering Problems. Theory of Approximation* (Academic, New York, 1976)

51. W.A. Cook, Body oriented (natural) co-ordinates for generating three-dimensional meshes. Int. J. Numer. Meth. Eng. **8**, 27–43 (1974)
52. W.J. Gordon, C.A. Hall, Construction of curvilinear co-ordinate systems and applications to mesh generation. Int. J. Numer. Meth. Eng. **7**, 461–477 (1973)
53. O.C. Zienkiewicz, D.V. Phillips, An automatic mesh generation scheme for plane and curved surfaces by 'isoparametric' co-ordinates. Int. J. Num. Methods Eng. **3**, 519–528 (1971)
54. H.D. Cohen, A method for the automatic generation of triangular elements on a surface. Int. J. Numer. Meth. Eng. **15**, 470–476 (1980)
55. W.D. Barfield, Numerical method for generating orthogonal curvilinear meshes. J. Comput. Phys. **5**, 23–33 (1970)
56. P.R. Brown, A non-interactive method for the automatic generation of finite element meshes using the Schwarz-Christoffel transformation. Comput. Methods Appl. Mech. Eng. **25**, 101–126 (1981)
57. K.H. Baldwin, H.L. Schreyer, Automatic generation of quadrilateral elements by a conformal mapping. Eng. Comput. **2**, 187–194 (1985)
58. A. Bykat, Design of a recursive, shape controlling mesh generator. Int. J. Numer. Meth. Eng. **19**, 1375–1390 (1983)
59. A. Bykat, Automatic generation of triangular grid: I—Subdivision of a general polygon into convex subregions. II—Triangulation of convex polygons. Int. J. Numer. Meth. Eng. **10**, 1329–1342 (1976)
60. M.L.C. Sluiter, D.C. Hansen, A general purpose automatic mesh generator for shell and solid finite elements. Comput. Eng., Vol. 3, Book No. G00217, ASME, pp. 29–34, (1982)
61. D.A. Lindholm, Automatic triangular mesh generation on surfaces of polyhedra. IEEE Trans. Mag. **MAG–19**(6), 1539–1542 (1983)
62. T.C. Woo, T. Thomasma, An algorithm for generating solid elements in objects with holes. Comput. Struct. **18**(2), 333–342 (1984)
63. M.A. Yerry, M.S. Shephard, Automatic three-dimensional mesh generation by the modified-octree technique. Int. J. Numer. Meth. Eng. **20**(11), 1965–1990 (1984)
64. A. Jain, *Modern Methods for Automatic FE Mesh Generation. Modern Methods for Automating Finite Element Mesh Generation* (The American Society of Civil Engineers, USA, 1986)
65. K. Ho-Le, Finite element mesh generation methods: A review and classification. Comput. Aided Des. **20**(1), 27–38 (1988)
66. International Meshing Roundtable, http://www.imr.sandia.gov/, last accessed on September 26, 2012
67. J. Sarrate, A. Huerta, Efficient unstructured quadrilateral mesh generation. Int. J. Numer. Meth. Eng. **49**, 1327–1350 (2000)
68. S.J. Owen, S. Saigal, H-Morph: An indirect approach to advancing front hex meshing. Int. J. Numer. Meth. Eng. **49**, 289–312 (2000)
69. S.J. Owen, Hex-dominant mesh generation using 3D constrained triangulation. Comput. Aided Des. **33**, 211–220 (2001)
70. S.J. Owen, M.L. Staten, S.A. Canann, S. Saigal, Q-Morph: An indirect approach to advancing front quad meshing. Int. J. Numer. Meth. Eng. **44**, 1314–1340 (1999)
71. Y. Lu, R. Gadh, T.J. Tautges, Feature based hex meshing methodology: Feature recognition and volume decomposition. Comput. Aided Des. **33**, 221–232 (2001)
72. X.Y. Li, S.H. Teng, A. Üngör, Simultaneous refinement and coarsening for adaptive meshing. Eng. Comput. **15**, 280–291 (1999)
73. M. Halpbern, Industrial requirements and practices in finite element meshing: A survey of trends. in *Proceedings of 6th International Meshing Roundtable, SAND97-2399*, Sandia National Laboratories, 1997
74. A. Sheffer, M. Bercovier, Hexahedral meshing of non-linear volumes using voronoi faces and edges. Int. J. Numer. Meth. Eng. **49**, 329–351 (2000)
75. M. Lai, S. Benzley, D. White, Automated hexahedral mesh generation by generalized multiple source to multiple target sweeping. Int. J. Numer. Meth. Eng. **49**, 261–375 (2000)

76. M.L. Staten, S.A. Canann, S.J. Owen, BMSweep: Locating interior nodes during sweeping. Eng. Comput. **15**, 212–218 (1999)

77. P. Knupp, Applications of mesh smoothing: Copy, morph, and sweep on unstructured quadrilateral meshes. Int. J. Numer. Meth. Eng. **45**, 37–45 (1999)

78. T.J. Tautges, The generation of hexahedral meshes for assembly geometry: Survey and progress. Int. J. Numer. Meth. Eng. **50**, 2617–2642 (2001)

79. G. Dhondt, Unstructured 20-node brick element meshing. Comput. Aided Des. **33**, 233–249 (2001)

80. G. Dhondt, A new automatic hexahedral mesher based on cutting. Int. J. Numer. Meth. Eng. **50**, 2109–2126 (2001)

81. T. Tautges, T. Blacker, S. Mitchell, The whisker weaving algorithm: A connectivity-based method for constructing all-hexahedral finite element meshes. Int. J. Numer. Meth. Eng. **39**, 3327–3349 (1996)

82. N.T. Folwell, S.A. Mitchell, Reliable whisker weaving via curve contraction. Eng. Comput. **15**, 292–302 (1999)

83. I. Yokota, *From Topological Geometry to Projective Geometry*. Modern Mathematics Press, (1993)

84. An Date, *Introduction to Data Base System* (Addison-Wesley Publishing Company, Third Edition, 1981)

Chapter 5
Dynamic Thermal Analysis

5.1 Introduction

The thermal behaviours of machine tools are more influential in machining accuracy than most of the other factors. According to Bryan [1], the thermal error caused by the thermal deformation takes 40–70 % out of all working errors in the field of precision manufacturing. Therefore, the thermal analysis is taken into consideration as an example to apply CBC substitution to finite element analysis.

Since the CBC substitution [2] provides a means to adjust element nodes dynamically, the thermal analysis of machine tools under actual operating conditions, using finite element method (FEM), can be implemented continuously even though the relative motions are taking place. The integrated nodal data management, based on the data structure discussed in Chap. 4, ensures that the nodal information associated with the finite element meshes is suitable for dynamic analysis and appropriate for data retrieval. The domain boundaries between machine components can also be extracted automatically.

The FEM, adopted in the thermal analysis, is a general technique for constructing approximated solutions to such kind of problems. This method involves dividing the physical system, or the domain of solution, into a finite number of small subregions, or subdomains, known as finite elements, and using variational concepts to construct an approximation of the solution over the collection of the finite elements. Each element is essentially a simple unit, the behaviour of which can be readily analysed. The features of the overall system are accommodated by using a large number of elements. One of the attractions of the FEM is the ease with which it can be applied to real engineering problems involving complex geometric features. The price that must be paid for flexibility and simplicity of individual elements is in the amount of numerical computations required to solve the resulting sets of simultaneous algebraic equations. Because of the generality and richness of the ideas underlying the method, it has been used with remarkable success in solving a wide range of problems in virtually all areas of engineering and mathematical physics.

L. Wang, *Dynamic Thermal Analysis of Machines in Running State*,
DOI: 10.1007/978-1-4471-5273-6_5, © Springer-Verlag London 2014

Although various efforts have been devoted to improve the popular FEM [3, 4], there still lacks a practical scheme for dynamic thermal analysis of machine tool, especially when it is put under actual operating or running state. Targeting this problem, this chapter introduces a new way to automate the process of dynamic thermal analysis with use of the CBC substitution and interpolation of intermediate results between consecutive FEA calculating steps.

This chapter is organised as follows. After the formulation of the thermal analysis, the models and conditions for FEA calculations are clarified. The method of thermal analysis based on an approach to interpolating the intermediate results is then described. A simplified two-component example is used to explain the concept and procedures for ease of understanding. Finally, discussions are provided to evaluate the approach. The contents included in this chapter are summarised based on the previous related research of the author [5–7].

5.2 Formulation of Thermal Analysis

5.2.1 General Considerations

The FEM was first applied to engineering problems of stiffness and deflection analysis of complex structures in the 1950s [8]. Later, in 1966, it was applied to some thermal problems such as heat conduction, temperature distribution and thermal deformation [9, 10].

When different parts of a rigid body are at different temperatures, heat flows from the hotter parts to the cooler ones. There are three distinct methods by which this transference of heat takes place:

1. Conduction, in which heat passes through the substance of the body itself;
2. Convection, in which heat is transferred by relative motion of portions of the heated body; and
3. Radiation, in which heat is transferred directly between distant portions of the body by electromagnetic radiation.

In liquids and gases, convection and radiation are of paramount importance, but in solids convection is altogether absent and radiation is usually negligible. In this chapter, therefore, only heat conduction is considered for the reasons mentioned above.

The thermal analysis of solids is generally carried out in the following way:

1. Determination of temperature distribution and
2. Calculation of thermal deformation based on the temperature distribution obtained.

Since the problem of temperature distribution is key for the entire thermal analysis, special attention is therefore given only to (1), the temperature distribution, at present for a case of three-dimensional nonstationary heat conduction with use of the FEM.

The fundamental concept of FEM is that any continuous field variable, such as temperature, can be approximated by a discrete model composed of a set of piecewise continuous field variables defined over a finite number of sub-domains—the elements. These elements are interconnected at specified joints called nodes or nodal points. Since the actual variation of the field variable inside the continuum is not known, some approximating functions are needed to describe its variation. These approximating functions, which are also known as the interpolating functions, are defined in terms of the values of the field variable at the nodal points. When the field equations, such as the equilibrium or the heat balance, for the whole body are written, the new unknowns will be the nodal values of the field variable. By solving the field equations, which are generally in the form of bounded matrices, the nodal value of the field variable can be obtained throughout the assemblage of elements [11–16].

The general solution of the thermal problem can be detailed in a step-by-step procedure. This sequence of steps describes the actual solution process which is followed in setting up and solving the heat conduction problems. In this chapter, the approach adopted is concentrated into a summary which is given below.

1. Idealisation of structure: the geometric features of the structure are simplified in order to accommodate sensible discretisation.
2. Discretisation of the structure: in this case, the body is subdivided into an equivalent system of finite elements. The type, size and number of elements are dictated by the geometric features of the component, the applied heat sources, the accuracy needed and the size of computer.
3. Introduction of governing equation of heat conduction: in this research, the heat conduction is considered as in a case of three-dimensional nonstationary one for an isotropic solid model. The appropriate differential equation of heat conduction for such kind of problem is introduced.
4. Discretisation of governing function: Ritz method [16], which is one of the direct methods using the variational principle, is adopted as an approximation on each individual element to introduce the finite element functions required.
5. Choice of interpolation or shape function: the assumed interpolation function approximates the actual or exact distribution of the temperature field within the continuum. In general, the interpolation function is taken in the form of a polynomial, and practical considerations limit the number of terms that can be retained in the polynomial.
6. Derivation of element heat conductivity matrix: the heat conductivity matrix is composed of the coefficients of the heat balance equations derived from the material and geometric properties of an element and obtained by the use of Fourier's law. The heat conductivity $[K]^e$ relates the temperatures at the nodal points $\{\theta\}^e$ to the applied heat flux at the nodal points $\{F\}^e$, where e denotes the element number.
7. Derivation of element heat capacity matrix: the heat capacity matrix $[C]^e$ is also composed of the coefficients of the heat balance equations derived from the

material and geometric properties of an element, which relates the differential of temperatures at the nodal points $\{\partial\theta/\partial t\}^e$ to $\{F\}^e$.

8. Assembly of element equations for the overall discretised body: this process includes the assembly of the global heat conductivity matrix $[K]$ for the entire body from the individual element heat conductivity matrices $[K]^e$, the global heat capacity matrix $[C]$ from the element heat capacity matrices $[C]^e$ and the global heat flux vector $\{F\}$ from the element nodal heat flux vectors $\{F\}^e$.

9. Solution for the unknown nodal temperatures: the overall heat balance equations have to be modified to account for the boundary conditions of the problem. After the incorporation of the boundary conditions, the global finite element equations can be expressed as

$$[K]\{\theta\} + [C]\left\{\frac{\partial\theta}{\partial t}\right\} = \{F\}$$

For problems with no relative motions, the temperature vector can be obtained easily. But for problems with relative motions between machine parts, the solution is obtained in a sequence of steps, each step involving the interpolating of temperature vector between steps, and the updating of the heat conductivity matrix $[K]$, heat capacity matrix $[C]$ and/or heat flux vector $\{F\}$.

The application of the above-mentioned nine steps of the finite element analysis is best demonstrated by an example to be discussed in Sect. 5.4.

5.2.2 Formulation

Governing Equation

The governing equation for nonstationary heat conduction in continuum (solid) can be expressed as follows [17]:

$$\rho c \frac{\partial T}{\partial t} = \frac{\partial}{\partial x}\left(\lambda_x \frac{\partial T}{\partial x}\right) + \frac{\partial}{\partial y}\left(\lambda_y \frac{\partial T}{\partial y}\right) + \frac{\partial}{\partial z}\left(\lambda_z \frac{\partial T}{\partial z}\right) + \dot{Q}, \qquad (5.1)$$

where ρ is the specific gravity, c the specific heat and λ_x, λ_y, λ_z the heat conductivities in directions x, y, z, respectively. Also, $T = T(x, y, z, t)$ is the temperature which is the function of space and time, and $\dot{Q} = \dot{Q}(x, y, z, t)$ is the rate of heat generated in a solid per unit time per unit volume.

Since only isotropic solids are taken into consideration in this chapter, which means that $\lambda_x = \lambda_y = \lambda_z = \lambda$ (constant), and $\dot{Q}$ is also assumed to be constant, Eq. (5.1) can be replaced by

$$\rho c \frac{\partial T}{\partial t} = \lambda\left(\frac{\partial^2 T}{\partial x^2} + \frac{\partial^2 T}{\partial y^2} + \frac{\partial^2 T}{\partial z^2}\right) + \dot{Q}, \qquad (5.2)$$

which is the governing function of the thermal analysis employed in this chapter.

Initial and Boundary Conditions

Before the approximation of the governing equation is carried out, it is necessary to determine the formulae which will express the initial and boundary conditions that the temperature satisfies. Here, it is assumed that in the interior of the solid T is a continuous function of x, y, z and t; and that this holds also for the first differential coefficient with regard to t and for the first and second differential coefficients with regard to x, y and z.

The initial condition of heat conduction is given by

$$T = T(x, y, z, 0),\tag{5.3}$$

at the instant being taken as the origin of the time coordinate t.

The boundary conditions usually arising in the mathematical theory of heat conduction are the following:

- *Prescribed temperature on surface S_1*. This temperature may be constant, or a function of time, or position, or both. The constant is considered here and is given as follows:

$$T = \bar{T} \quad \text{(on the surface } S_1)\tag{5.4}$$

where $\bar{T}$ is the prescribed temperature on S_1.
- *Prescribed heat flux across surface S_2*. That is,

$$q = -\lambda\frac{\partial T}{\partial n} = q_0 \quad \text{(on the surface } S_2)\tag{5.5}$$

where n is an outward normal to S_2, and q_0 the heat flux across S_2.
- *Heat convection on surface S_3 which is given by*

$$q = \alpha_c(T - T_c) \quad \text{(on the surface } S_3)\tag{5.6}$$

where α_c is heat transfer coefficient across S_3, and T_c the surrounding temperature.
- *Radiation boundary condition at surface S_4*. If the heat flux across the surface is proportional to the temperature difference between the surface and the surrounding medium, it is given by

$$\begin{aligned} q &= \varepsilon\sigma F(T^4 - T_r^4) \\ &= \alpha_r(T - T_r) \quad \text{(on the surface } S_4)\end{aligned}\tag{5.7}$$

where ε is the rate of heat radiation, σ the Stefan–Blotzmann constant, F the shape coefficient and T_r the temperature of radiation source, respectively.

Approximation to Finite Element Functions

In this chapter, the Ritz method [16] is adopted for finding the approximate solution of Governing equation (5.2) together with the boundary conditions (5.5–5.7), although the Galerkin method [16] is also applicable to problems of this type.

Each method to obtain an approximate solution of a boundary value problem requires some kind of discretisation. In the case of the Ritz method, this is done by minimising the functional I given by (5.8) in which $T(x, y, z, t)$ satisfies Eq. (5.2) and conditions (5.5–5.7).

$$I = \int_v \left[\frac{1}{2}\lambda \left\{ \left(\frac{\partial T}{\partial x}\right)^2 + \left(\frac{\partial T}{\partial y}\right)^2 + \left(\frac{\partial T}{\partial z}\right)^2 \right\} - \left(\dot{Q} - \rho c \frac{\partial T}{\partial t} \right) T \right] dv$$

$$+ \int_{S_2} q_0 T \, dS + \int_{S_3} \frac{1}{2}\alpha_c \left(T^2 - 2T_c T \right) dS + \int_{S_4} \frac{1}{2}\alpha_r \left(T^2 - 2T_r T \right) dS \quad (5.8)$$

Since $T(x, y, z, t)$ in Eq. (5.8) expresses a temperature set of infinite points within a solid, it will introduce a simple equation system containing infinite equations if the condition $\partial I/\partial T = 0$ is applied to the case.

In order to utilise the thermal analysis for FEM, functional I is discretised to individual finite elements. Therefore, I can be expressed as the sum of I^e as follows:

$$I = \sum_{e=1}^{k} I^e \quad (5.9)$$

$$I^e = \int_{v^e} \left[\frac{1}{2}\lambda \left\{ \left(\frac{\partial T}{\partial x}\right)^2 + \left(\frac{\partial T}{\partial y}\right)^2 + \left(\frac{\partial T}{\partial z}\right)^2 \right\} - \left(\dot{Q} - \rho c \frac{\partial T}{\partial t} \right) T \right] dv$$

$$+ \int_{S_2^e} q_0 T \, dS + \int_{S_3^e} \frac{1}{2}\alpha_c \left(T^2 - 2T_c T \right) dS$$

$$+ \int_{S_4^e} \frac{1}{2}\alpha_r \left(T^2 - 2T_r T \right) dS \quad (5.10)$$

where k is the number of elements, and I^e the functional of element e.

Furthermore, the temperature distribution within the interior of the elements is given by

$$T(x, y, z, t) = \left[N(x, y, z) \right] \{\theta(t)\} \quad (5.11)$$

where $\left[N(x, y, z) \right]$ is the shape function matrix which is related only to the shape of element, the number of nodes and the nodal layout of element; and $\{\theta(t)\}$ is the nodal temperature vector at time t.

Substituting Eqs. (5.11) into (5.10), functional I^e may be replaced by

$$
\begin{aligned}
I^e = \{\theta(t)\}^T \int_{v^e} \Bigg[&\frac{1}{2}\lambda \left(\frac{\partial[N]^T}{\partial x}\frac{\partial[N]}{\partial x} + \frac{\partial[N]^T}{\partial y}\frac{\partial[N]}{\partial y} \right. \\
&+ \left. \frac{\partial[N]^T}{\partial z}\frac{\partial[N]}{\partial z} \right) \{\theta(t)\} - [N]^T \left(\dot{Q} - \rho c[N]\frac{\partial\{\theta(t)\}}{\partial t} \right) \Bigg] dv \\
&+ \{\theta(t)\}^T \int_{S_2^e} [N]^T q_0 dS \\
&+ \{\theta(t)\}^T \int_{S_3^e} \frac{1}{2}\alpha_c[N]^T \left([N]\{\theta(t)\} - 2T_c \right) dS \\
&+ \{\theta(t)\}^T \int_{S_4^e} \frac{1}{2}\alpha_r[N]^T \left([N]\{\theta(t)\} - 2T_r \right) dS
\end{aligned}
\tag{5.12}
$$

where, $\{\ \}^T$ and $[\ ]^T$ denote the transpositional forms of $\{\ \}$ and $[\ ]$, respectively. Then, differentiating Eq. (5.12) with respect to the nodal temperature θ gives

$$
\begin{aligned}
\frac{\partial I^e}{\partial\theta} = &\frac{\partial\{\theta(t)\}^T}{\partial\theta} \int_{v^e} \lambda \left(\frac{\partial[N]^T}{\partial x}\frac{\partial[N]}{\partial x} + \frac{\partial[N]^T}{\partial y}\frac{\partial[N]}{\partial y} + \frac{\partial[N]^T}{\partial z}\frac{\partial[N]}{\partial z} \right) dv \cdot \{\theta(t)\} \\
&- \frac{\partial\{\theta(t)\}^T}{\partial\theta} \int_{v^e} \dot{Q}[N]^T dv + \frac{\partial\{\theta(t)\}^T}{\partial\theta} \int_{v^e} \rho c[N]^T[N]dv \cdot \left\{ \frac{\partial\theta(t)}{\partial t} \right\} \\
&+ \frac{\partial\{\theta(t)\}^T}{\partial\theta} \int_{S_2^e} q_0[N]^T dS \\
&+ \frac{\partial\{\theta(t)\}^T}{\partial\theta} \int_{S_3^e} \alpha_c[N]^T[N]dS \cdot \{\theta(t)\} - \frac{\partial\{\theta(t)\}^T}{\partial\theta} \int_{S_3^e} \alpha_c T_c[N]^T dS \\
&+ \frac{\partial\{\theta(t)\}^T}{\partial\theta} \int_{S_4^e} \alpha_r[N]^T[N]dS \cdot \{\theta(t)\} - \frac{\partial\{\theta(t)\}^T}{\partial\theta} \int_{S_4^e} \alpha_r T_r[N]^T dS.
\end{aligned}
\tag{5.13}
$$

According to the variational method, the finite element functions for the problems of three-dimensional nonstationary heat conduction can be obtained as follows by setting $\partial I^e/\partial\theta = 0$

$$
[K]^e\{\theta\} + [C]^e \left\{ \frac{\partial\theta}{\partial t} \right\} = \{F\}^e
\tag{5.14}
$$

where,

$$
[K]^e = \int_{v^e} \lambda \left(\frac{\partial[N]^T}{\partial x}\frac{\partial[N]}{\partial x} + \frac{\partial[N]^T}{\partial y}\frac{\partial[N]}{\partial y} + \frac{\partial[N]^T}{\partial z}\frac{\partial[N]}{\partial z} \right) dv
$$

$$+ \int_{S_3^e} \alpha_c [N]^T [N] dS + \int_{S_4^e} \alpha_r [N]^T [N] dS \qquad (5.15)$$

$$[C]^e = \int_{v^e} \rho c [N]^T [N] dv \qquad (5.16)$$

$$\{F\}^e = \int_{v^e} \dot{Q} [N]^T dv - \int_{S_2^e} q_0 [N]^T dS$$

$$+ \int_{S_3^e} \alpha_c T_c [N]^T dS + \int_{S_4^e} \alpha_r T_r [N]^T dS \qquad (5.17)$$

Assembling Eq. (5.14) of element $e (e \in [1, k])$, the finite element function for all the elements is obtained as

$$[K]\{\theta\} + [C] \left\{ \frac{\partial \theta}{\partial t} \right\} = \{F\} \qquad (5.18)$$

where,

$$[K] = \sum_{e=1}^{k} [K]^e : \text{heat conductivity matrix} \qquad (5.19)$$

$$[C] = \sum_{e=1}^{k} [C]^e : \text{heat capacity matrix} \qquad (5.20)$$

$$\{F\} = \sum_{e=1}^{k} \{F\}^e : \text{heat flux vector.} \qquad (5.21)$$

The nonstationary heat conduction depends on not only the space but also on the time. Until now, only discretisation for the space has been carried out. In order that the discretisation for the time can be carried out easily, Crank-Nicolson's finite difference functions [18] are adopted. According to this method, nodal temperature vector at time $t + \Delta t / 2$ can be expressed as

$$\left\{ \theta \left(t + \frac{\Delta t}{2} \right) \right\} = \frac{1}{2} (\{\theta(t + \Delta t)\} + \{\theta(t)\}) \qquad (5.22)$$

and the differentiation of nodal temperature vector with respect to time t at $t + \Delta t / 2$ can be expressed as

$$\left\{ \frac{\partial \theta}{\partial t} \left(t + \frac{\Delta t}{2} \right) \right\} = \frac{\{\theta(t + \Delta t)\} - \{\theta(t)\}}{\Delta t} \qquad (5.23)$$

Substituting Eqs. (5.22) and (5.23) into (5.18), the finite element function is replaced by

$$[K]\left\{\frac{1}{2}\big(\{\theta(t+\Delta t)\}+\{\theta(t)\}\big)\right\}+[C]\left\{\frac{1}{\Delta t}\big(\{\theta(t+\Delta t)\}-\{\theta(t)\}\big)\right\}=\{F\} \quad (5.24)$$

and the final finite element function for the nonstationary heat conduction becomes

$$\left(\frac{1}{2}[K]+\frac{1}{\Delta t}[C]\right)\{\theta(t+\Delta t)\}=\left(-\frac{1}{2}[K]+\frac{1}{\Delta t}[C]\right)\{\theta(t)\}+\{F\} \quad (5.25)$$

If $\{\theta(t)\}$ in Eq. (5.25) is known, the problem of nonstationary heat conduction can be solved by following the calculating steps described above.

5.3 Temperature Distribution Analysis

5.3.1 Model and Conditions

When a machine tool is in running state, its relative motions, such as feed motion and cutting motion, are required in order to generate a pre-defined shape out of a workpiece. The table and base among the components of the machine tool are extracted and considered as a two-component example in this section. The extracted pair of the table and base is simplified and modelled as shown in the left of Fig. 5.1. A heat source is embedded on the top surface of the table. Since the main objective of the thermal analysis is to prove that the CBC substitution is applicable and suitable for dynamic analysis of objects with relative motions, the simplified table and base together are considered as the model for analysis. The actual sizes are scaled down in this model.

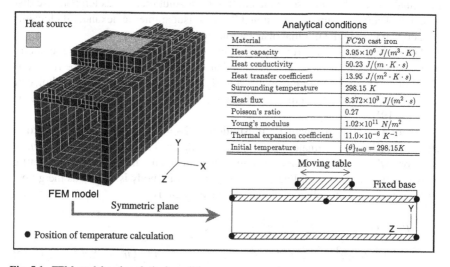

Analytical conditions	
Material	$FC20$ cast iron
Heat capacity	$3.95\times10^6\ J/(m^3\cdot K)$
Heat conductivity	$50.23\ J/(m\cdot K\cdot s)$
Heat transfer coefficient	$13.95\ J/(m^2\cdot K\cdot s)$
Surrounding temperature	$298.15\ K$
Heat flux	$8.372\times10^3\ J/(m^2\cdot s)$
Poisson's ratio	0.27
Young's modulus	$1.02\times10^{11}\ N/m^2$
Thermal expansion coefficient	$11.0\times10^{-6}\ K^{-1}$
Initial temperature	$\{\theta\}_{t=0}=298.15K$

Heat source

FEM model

Symmetric plane

● Position of temperature calculation

Moving table

Fixed base

Fig. 5.1 FEM model and analytical conditions

Table 5.1 Conditions for temperature distribution analysis

Material	$FC20$ cast iron
Heat capacity	$3.95 \times 10^6 \, \text{J}/(\text{m}^3 \cdot \text{K})$
Heat conductivity	$50.23 \, \text{J}/(\text{m} \cdot \text{K} \cdot \text{s})$
Heat transfer coefficient	$13.95 \, \text{J}/(\text{m}^2 \cdot \text{K} \cdot \text{s})$
Surrounding temperature	$298.15 \, \text{K}$
Heat flux	$8.372 \times 10^3 \, \text{J}/(\text{m}^2 \cdot \text{s})$
Poisson's ratio	0.27
Young's modulus	$1.02 \times 10^{11} \, \text{N}/\text{m}^2$
Thermal expansion coefficient	$11.0 \times 10^{-6} \, \text{K}^{-1}$
Initial temperature	$\{\theta\}_{t=0} = 298.15 \, \text{K}$

The material of the model chosen is $FC20$ cast iron. The initial temperature of the entire model is set to 298.15 K. The surrounding temperature is also set to 298.15 K. The other conditions necessary for the temperature distribution analysis are summarised in the upper-right of Fig. 5.1 and in Table 5.1.

5.3.2 Shape Functions of Isoparametric Elements

As mentioned in Chap. 4, the finite elements used in this book are hexahedra. A general hexahedron has eight primary external nodes. In this case, the linear element is restricted to plane surfaces. But isoparametric elements, which belong to higher order elements than the linear one, can have curved surfaces for their sides. This fact is of importance not only to the body with curved surfaces which will be divided into small hexahedra but also to the continuity of solution such as temperature distribution. Therefore, three-dimensional 20-node isoparametric hexahedral elements are adopted here in the thermal distribution analysis.

The shape functions approximate the actual or exact distribution of the temperature field within an element. It is evident that the choice of suitable shape functions plays an important role in finite element analysis, because the appropriate choice of shape functions leads to elements of high accuracy and with converging characteristics [4, 19].

Figure 5.2 shows a schematic illustration of 20-node isoparametric hexahedral element in a local coordinate system (ξ, η, ζ). The local coordinate system (ξ, η, ζ) is the one which is defined for a particular element and not necessarily for the entire body or structure. The coordinate system for the entire body is known as the global system (x, y, z). The polynomial basis of the element is

$$\langle P \rangle = \langle 1 \ \xi \ \eta \ \zeta; \ \xi^2 \ \xi\eta \ \eta^2 \ \eta\zeta \ \zeta^2 \ \xi\zeta;$$
$$\xi^2 \ \eta \ \xi\eta^2 \ \eta^2 \ \zeta \ \eta\zeta^2 \ \xi\zeta^2 \ \xi^2 \ \zeta \ \xi\eta\zeta; \ \xi^2 \ \eta\zeta \ \xi\eta^2 \ \zeta \ \xi\eta\zeta^2 \rangle \quad (5.26)$$

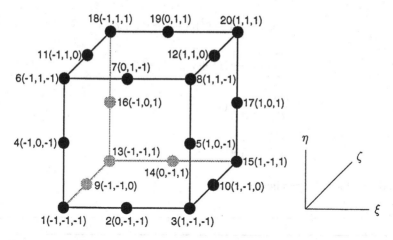

Fig. 5.2 Nodal layout of a 20-node isoparametric element

Shape functions N_i and their derivatives are given below:

- Corner nodes

Node i	1	3	6	8	13	15	18	20
ξ_i	-1	1	-1	1	-1	1	-1	1
η_i	-1	-1	1	1	-1	-1	1	1
ζ_i	-1	-1	-1	-1	1	1	1	1

$$N_i = \frac{1}{8}(1 + \xi\xi_i)(1 + \eta\eta_i)(1 + \zeta\zeta_i)(-2 + \xi\xi_i + \eta\eta_i + \zeta\zeta_i)$$

$$\frac{\partial N_i}{\partial \xi} = \frac{1}{8}\xi_i(1 + \eta\eta_i)(1 + \zeta\zeta_i)(-1 + 2\xi\xi_i + \eta\eta_i + \zeta\zeta_i)$$

$$\frac{\partial N_i}{\partial \eta} = \frac{1}{8}\eta_i(1 + \xi\xi_i)(1 + \zeta\zeta_i)(-1 + \xi\xi_i + 2\eta\eta_i + \zeta\zeta_i)$$

$$\frac{\partial N_i}{\partial \zeta} = \frac{1}{8}\zeta_i(1 + \xi\xi_i)(1 + \eta\eta_i)(-1 + \xi\xi_i + \eta\eta_i + 2\zeta\zeta_i)$$

- Nodes on the sides parallel to axis ξ

Node i	2	7	14	19
$\xi_i = 0;\ \eta_i$	-1	1	-1	1
ζ_i	-1	-1	1	1

$$N_i = \frac{1}{4}(1 - \xi^2)(1 + \eta\eta_i)(1 + \zeta\zeta_i)$$

$$\frac{\partial N_i}{\partial \xi} = -\frac{1}{2}\xi(1 + \eta\eta_i)(1 + \zeta\zeta_i)$$

$$\frac{\partial N_i}{\partial \eta} = \frac{1}{4}\eta_i(1 - \xi^2)(1 + \zeta\zeta_i)$$

$$\frac{\partial N_i}{\partial \zeta} = \frac{1}{4}\zeta_i(1 - \xi^2)(1 + \eta\eta_i)$$

- Nodes on the sides parallel to axis η

Node i	4	5	16	17
$\eta_i = 0; \xi_i$	-1	1	-1	1
ζ_i	-1	-1	1	1

$$N_i = \frac{1}{4}(1 + \xi\xi_i)(1 - \eta^2)(1 + \zeta\zeta_i)$$

$$\frac{\partial N_i}{\partial \xi} = \frac{1}{4}\xi_i(1 - \eta^2)(1 + \zeta\zeta_i)$$

$$\frac{\partial N_i}{\partial \eta} = -\frac{1}{2}\eta(1 + \xi\xi_i)(1 + \zeta\zeta_i)$$

$$\frac{\partial N_i}{\partial \zeta} = \frac{1}{4}\zeta_i(1 - \xi\xi_i)(1 - \eta^2)$$

- Nodes on the sides parallel to axis ζ

Node i	9	10	11	12
$\zeta_i = 0; \xi_i$	-1	1	-1	1
η_i	-1	-1	1	1

$$N_i = \frac{1}{4}(1 + \xi\xi_i)(1 + \eta\eta_i)(1 - \zeta^2)$$

$$\frac{\partial N_i}{\partial \xi} = \frac{1}{4}\xi_i(1 + \eta\eta_i)(1 - \zeta^2)$$

$$\frac{\partial N_i}{\partial \eta} = \frac{1}{4}\eta_i(1 + \xi\xi_i)(1 - \zeta^2)$$

$$\frac{\partial N_i}{\partial \zeta} = -\frac{1}{2}\zeta(1 + \xi\xi_i)(1 + \eta\eta)$$

The coordinate transformation from the local coordinate system (ξ, η, ζ) to the global coordinate system (x, y, z) can be expressed as

$$x(\xi, \eta, \zeta) = N'_i(\xi, \eta, \zeta)x_i$$
$$y(\xi, \eta, \zeta) = N'_i(\xi, \eta, \zeta)y_i \qquad (5.27)$$
$$z(\xi, \eta, \zeta) = N'_i(\xi, \eta, \zeta)z_i$$

where, $-1 \leq \xi, \eta, \zeta \leq 1$ and (x_i, y_i, z_i) denote the ith node within an element in the global coordinate system. Since the elements used in this research are isoparametric elements, their transformation matrices are equal to their shape functions [20].

$$N'_i = N_i \qquad (5.28)$$

This is useful for relating the coordinates of elements in both the local and the global coordinate systems.

5.3.3 Finite Element Calculation

Based on the shape functions mentioned above, matrices $[K]^e$, $[C]^e$ and vector $\{F\}^e$ for each individual element e can be calculated using Eqs. (5.15–5.17). However, the shape functions for isoparametric elements are defined in terms of the local coordinates ξ, η, ζ, and therefore they cannot be differentiated directly with respect to the global coordinates x, y, z.

In order to overcome this difficulty, it is necessary to obtain a relationship between the derivatives of the two sets of coordinate systems. The partial derivatives of the shape function $N_i(\xi, \eta, \zeta)$ with respect to ξ, η, ζ can be expressed as

$$\begin{Bmatrix} \dfrac{\partial N_i}{\partial \xi} \\[2mm] \dfrac{\partial N_i}{\partial \eta} \\[2mm] \dfrac{\partial N_i}{\partial \zeta} \end{Bmatrix} = \begin{bmatrix} \dfrac{\partial x}{\partial \xi} & \dfrac{\partial y}{\partial \xi} & \dfrac{\partial z}{\partial \xi} \\[2mm] \dfrac{\partial x}{\partial \eta} & \dfrac{\partial y}{\partial \eta} & \dfrac{\partial z}{\partial \eta} \\[2mm] \dfrac{\partial x}{\partial \zeta} & \dfrac{\partial y}{\partial \zeta} & \dfrac{\partial z}{\partial \zeta} \end{bmatrix} \begin{Bmatrix} \dfrac{\partial N_i}{\partial x} \\[2mm] \dfrac{\partial N_i}{\partial y} \\[2mm] \dfrac{\partial N_i}{\partial z} \end{Bmatrix} = [J] \begin{Bmatrix} \dfrac{\partial N_i}{\partial x} \\[2mm] \dfrac{\partial N_i}{\partial y} \\[2mm] \dfrac{\partial N_i}{\partial z} \end{Bmatrix} \qquad (5.29)$$

where, $[J]$ is the Jacobian matrix. According to Eq. (5.27), x, y, z can be described as the functions of ξ, η, ζ. Therefore, $[J]$ can also be expressed as the functions of ξ, η, ζ

$$[J] = \begin{bmatrix} \sum \frac{\partial N_i}{\partial \xi} x_i & \sum \frac{\partial N_i}{\partial \xi} y_i & \sum \frac{\partial N_i}{\partial \xi} z_i \\ \sum \frac{\partial N_i}{\partial \eta} x_i & \sum \frac{\partial N_i}{\partial \eta} y_i & \sum \frac{\partial N_i}{\partial \eta} z_i \\ \sum \frac{\partial N_i}{\partial \zeta} x_i & \sum \frac{\partial N_i}{\partial \zeta} y_i & \sum \frac{\partial N_i}{\partial \zeta} z_i \end{bmatrix}$$

$$= \begin{bmatrix} \frac{\partial N_1}{\partial \xi} & \frac{\partial N_2}{\partial \xi} & \cdots \\ \frac{\partial N_1}{\partial \eta} & \frac{\partial N_2}{\partial \eta} & \cdots \\ \frac{\partial N_1}{\partial \zeta} & \frac{\partial N_2}{\partial \zeta} & \cdots \end{bmatrix} \begin{bmatrix} x_1 & y_1 & z_1 \\ x_2 & y_2 & z_2 \\ \vdots & \vdots & \vdots \end{bmatrix} \qquad (5.30)$$

From Eq. (5.29), the partial derivatives of shape function with respect to x, y, z become

$$\begin{Bmatrix} \frac{\partial N_i}{\partial x} \\ \frac{\partial N_i}{\partial y} \\ \frac{\partial N_i}{\partial z} \end{Bmatrix} = [J]^{-1} \begin{Bmatrix} \frac{\partial N_i}{\partial \xi} \\ \frac{\partial N_i}{\partial \eta} \\ \frac{\partial N_i}{\partial \zeta} \end{Bmatrix} \qquad (5.31)$$

where $[J]^{-1}$ denotes the inverse matrix of $[J]$. This inverse exists if there is no excessive distortion of the element.

The integrals included in $[K]^e$, $[C]^e$ and $\{F\}^e$ are commonly evaluated numerically by using Gaussian quadrature method [20]. In the process of solving the temperature distribution problem, derivation of heat conductivity matrix $[K]$ remains a key point, because $[K]$ is often large, sparse and positive definite. If the bandwidth of $[K]$ is large, it will increase both the computing time and the storage requirement. It is usually required that the bandwidth of $[K]$ is as narrow as possible. This is done by adopting an automatic renumbering method [21]. At the mesh generation stage, the nodal number is allowed to be input in any form. Users pay no attention to how the nodes should be numbered and how the bandwidth of matrix could be reduced. These tasks are implemented internally in the integrated CAD/CAE system.

The temperature distribution calculations are carried out under several sets of conditions within 30 min. First, the temperature distributions of the model defined above are calculated when its upper block (table) is fixed at the left, the middle and the right of the lower block (base), respectively. Since the model is symmetric in x direction, only the symmetric plane of the model shown in the lower-right of Fig. 5.1 is extracted and seven specific points are defined on the plane. In this case, since no motion exists between the table and the base, they are equivalent to one rigid body. The temperature rises steadily over time. However, if the table moves against the base during FEA calculation, special treatments of the intermediate results are needed at the contacting boundary of the two components. An interpolation method is introduced below.

5.3.4 Interpolation for Continuous Calculation

When a machine tool is in running state, its table moves against the base regularly. Assume that the same situation occurs during thermal analysis. When the relative motion is present, the nodes of elements on the contacting surface will disagree. Meanwhile, the node adjustment is totally controlled by the CBC substitution as mentioned in Chap. 4. A dynamically updated FEM model can be obtained whenever a relative motion takes place. What remains to be done is how to deal with the problem of interpolation of nodal temperature values among consecutive calculating steps, since the nodal positions of two adjacent steps near the contacting surface may be different.

The overall procedures of nodal value interpolation are summarised as follows:

1. Calculate the temperature distribution of the model at time t and record all nodal temperature values in a buffer including temperature T_t and its incremental rate $\dot{T}_t$. These temperature values will be used as the basis for the interpolation.
2. Adjust nodal connectivity when the relative position of the table and base is changed after a time interval Δt by CBC substitution, and recalculate those nodal coordinates accordingly.
3. Since the nodal positions at time $t + \Delta t$ are known, the new nodal temperature values are derived based on T_t and $\dot{T}_t$. The shape functions of isoparametric hexahedra are chosen as the interpolation functions.
4. Recalculate the matrices $[K]$, $[C]$, and $\{F\}$, since the total number of elements may be different due to the relative motion.
5. Derive new temperature distribution $T_{t+\Delta t}$ and $\dot{T}_{t+\Delta t}$ at time $t + \Delta t$ by Eq. (5.25).

The above procedures are repeated whenever a relative motion occurs until the thermal analysis is completed.

In addition to the shape functions, other polynomial type of interpolation functions are also suitable for nodal temperature interpolation. However, the following pertinent points must be taken into account in deciding upon the order of the polynomial. These are:

- The interpolation function must satisfy the convergence requirements [12];
- The number of generalised coordinates should be equal to the number of the nodal degrees of freedom of the element; and
- The pattern of variation of the field variable, such as temperature, should be independent of the local coordinate system. This property is known as geometric invariance or spatial isotropy. In order to achieve this, the polynomial should contain terms which do not violate the symmetry.

The detailed interpolation procedures are divided into primary interpolation and secondary interpolation. Assume that V^{old} is the nodal set at time t before nodal adjustment, and V^{new} the nodal set after the adjustment. The primary interpolation is first performed within a limited area not greater than $\pm 1.5 \delta^{\text{max}}$ to prevent any ill case in accuracy, where δ^{max} is the maximum mesh size of the current meshwork.

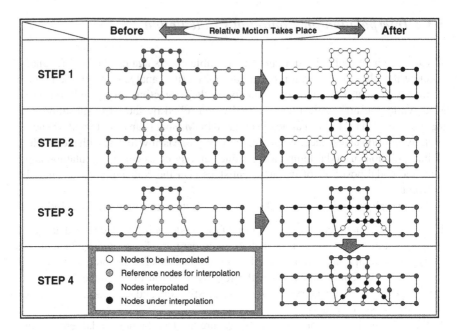

Fig. 5.3 A 2D example of analytical data interpolation

Linear interpolation is considered at this stage due to the fact of low gradient of temperature rise during Δt. The temperature data of each node is interpolated along x, y and z directions, respectively, and an average of the three temperature elements is taken as the new temperature data of the node. Since the primary interpolation cannot cover all the cases of element topology, a secondary interpolation is needed to cope with the remaining nodes, which is performed based on the results of the primary interpolation.

The above-mentioned interpolation includes four steps, the algorithms of which are described in C-like language. Figure 5.3 schematically illustrates a two-dimensional example for analytical data interpolation. The left column shows the meshwork at $t - \Delta t$ before a relative motion takes place, and the right column shows that at t after the relative motion occurred.

Primary Interpolation

The operation procedures of the primary interpolation are to judge the status of each individual node near the contacting surface of the two components by checking its nodal coordinates and hence to determine whether the node should be interpolated. The interpolation is done based on the information about its surrounding nodes.

for $\forall V_i^{\text{new}} (V_i^{\text{new}} \in (V^{\text{new}} \ominus {}^+V^{\text{new}}))$ do

    $status \Longleftarrow$ **Check_Nodal_Coordinate** $(V_i^{\text{new}}, V^{\text{old}})$;

    step_switch $(status) \rightsquigarrow$ (See Fig. 5.3 for reference)

      step 1: $d(V_i^{\text{new}}, V_j^{\text{old}}(V_j^{\text{old}} \in V^{\text{old}})) \equiv 0$

        $T_i^{\text{new}} = T_j^{\text{old}}$;

        $\dot{T}_i^{\text{new}} = \dot{T}_j^{\text{old}}$;

        break;

      step 2: $d(V_i^{\text{new}}, V_j^{\text{old}})_P \equiv \tilde{D} \wedge d(V_i^{\text{new}}, V_j^{\text{old}})_{\bar{P}} \equiv 0$

        $T_i^{\text{new}} = T_j^{\text{old}}$;

        $\dot{T}_i^{\text{new}} = \dot{T}_j^{\text{old}}$;

        break;

      step 3: $d(V_i^{\text{new}}, \forall V_j^{\text{old}}(V_j^{\text{old}} \in V^{\text{old}})) \not\equiv 0$

        for $\forall \omega (\omega \in \{x, y, z\})$ do

          if **Around** $(V_i^{\text{new}})_\omega \exists V_{m,n}^{\text{old}} \in V^{\text{old}} \wedge$

          $-1.5\delta^{\max} \le d(V_i^{\text{new}}, V_{m,n}^{\text{old}}) \le 1.5\delta^{\max}$

          Interpolate $(T_m^{\text{old}}, T_i^{\text{new}}, T_n^{\text{old}})_\omega$;

          Interpolate $(\dot{T}_m^{\text{old}}, \dot{T}_i^{\text{new}}, \dot{T}_n^{\text{old}})_\omega$;

        endif

        endfor

        Average $(T_i^{\text{new}})_{\omega \in \{x,y,z\}}$;

        Average $(\dot{T}_i^{\text{new}})_{\omega \in \{x,y,z\}}$;

        break;

    endswitch

    push $V_i^{\text{new}} \rightarrow {}^+V^{\text{new}} \subset V^{\text{new}}$;

endfor

where, ${}^+V^{\text{new}}$ is an adjusted nodal set with interpolated analytical data, $\tilde{D}$ the displacement of relative motion after time interval Δt, P the direction of the relative motion, $\bar{P}$ the direction perpendicular to P and ω the values corresponding to x, y and z terms, respectively.

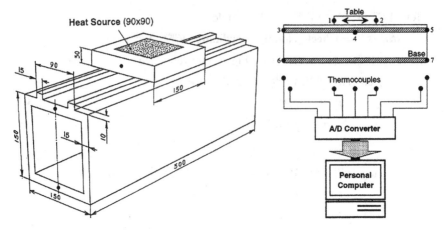

Fig. 5.4 Experimental setup for temperature measurement

Secondary Interpolation

The secondary interpolation is needed only if there are remaining nodes that cannot be covered and interpolated by the primary interpolation. The procedures for the secondary interpolation are given below.

$$\text{step 4: for } \forall V_i^{\text{new}} (V_i^{\text{new}} \not\in {}^+V^{\text{new}}) \text{ do}$$

$$\text{if } \textbf{Around} \, (V_i^{\text{new}}) \, \exists \, V_{m,n}^{\text{new}} \in {}^+V^{\text{new}} \wedge$$

$$-\delta^{\max} \leq d(V_i^{\text{new}}, V_{m,n}^{\text{new}}) \leq \delta^{\max}$$

$$\textbf{Interpolate} \, (T_m^{\text{new}}, T_i^{\text{new}}, T_n^{\text{new}});$$

$$\textbf{Interpolate} \, (\dot{T}_m^{\text{new}}, \dot{T}_i^{\text{new}}, \dot{T}_n^{\text{new}});$$

$$\text{endif}$$

$$\text{endfor}$$

$$\text{push } V_i^{\text{new}} \rightarrow {}^+V^{\text{new}} \subset V^{\text{new}};$$

$$\text{break;}$$

$$\text{endstep4}$$

For more details of the interpolation algorithms, interested readers are referred to [22].

5.4 A Case Study

The cutting heat generated between a workpiece and a cutter is transmitted to the base of a machine tool through the workpiece and the table. The thermal behaviour

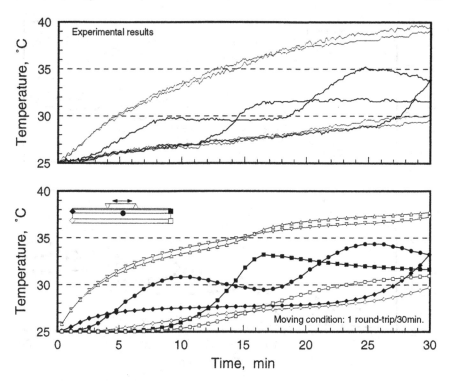

Fig. 5.5 Temperature rise (1 round-trip/30 min)

between the table and the base, which form a typical movable joint, is of vital importance in terms of assuring machining accuracy. Therefore, the simplified two-component example (a table moving on top of a fixed base), shown in Fig. 5.1, is chosen in this case study to validate the processing algorithms. The case study is to investigate and understand the unsteady-state (or running-state) heat conduction and thermal deformation of the table-base composition. A heat source, equivalent to the cutting heat, is fixed on the upper surface of the table. When the table moves relatively over the base together with the heat source, heat conduction and heat convection take place. Seven specific points in the symmetric plane parallel to Y-Z plane (see the lower-right illustration of Fig. 5.1) are defined and used to represent the temperature distribution in the case study.

In order to validate the analytical results and the processing algorithms, a set of experiments are conducted under the same conditions. The temperature data at the seven specific points are measured using thermocouples (ϕ 0.3 mm, chromel-alumel). Figure 5.4 gives the dimensions of test piece (left) and the connection of the thermocouples to a personal computer via an A/D converter (right).

Based on the procedures introduced previously, the temperature distribution analyses with relative motion are carried out under varying conditions. Corresponding experimental results are also collected. The window of the total time (to virtually

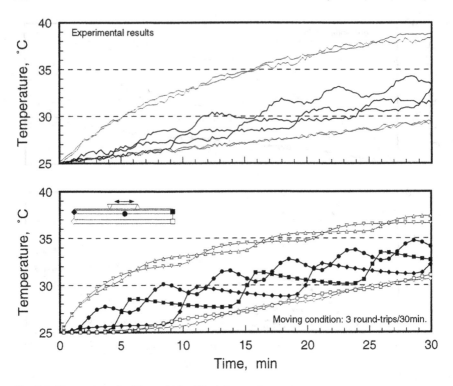

Fig. 5.6 Temperature rise (3 round-trips/30 min)

simulate and physically observe the thermal behaviours of the table and the base) is set to 30 min. In other words, the calculations and experiments start when the heat source is applied and the table moves at $t = 0$, and stop when $t = 30$ min. The calculations and experiments are repeated under different moving speeds of the table. Note that the time of 30 min is not the computation time but the simulated/experimental time for heat conduction, etc., of the table-base components in running state in the real world.

The results of the case study are shown in Figs. 5.5 and 5.6, each of which consists of the FEA calculation results (bottom) the experimental results (top). Figure 5.5 shows the temperature changes of the specific points when the table is moving slowly (1 round-trip/30 min) from the left to the right on the upper surface of the base and back to the left edge. Since the heat source is fixed on the table, the incremental temperature changes in the specific points are different due to the relative motion. This is true especially for the specific points affixed on the upper part of the base (the three dark markers in the figure). Figure 5.6 shows the temperature changes when the table moves faster at a rate of 3 round-trips/30 min. It is evident that the simulation results match the practical situations very well.

Finally, the results of temperature distribution and thermal deformation of the table-base model are illustrated in Fig. 5.7. The figure provides an overview of the

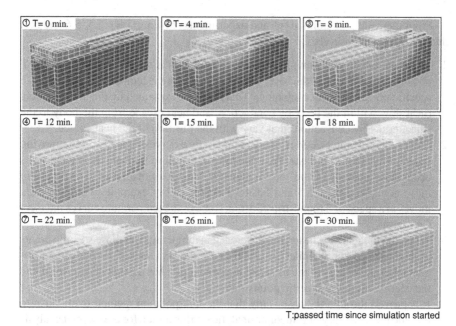

T:passed time since simulation started

Fig. 5.7 Temperature distribution and deformation with moving heat source

thermal behaviours of the model, intuitively. Because of the uneven temperature distribution, the table-base structure deforms accordingly over time. For better visibility, the thermal deformation is enlarged by a scale of 1:500. The results derived by the dynamic thermal analysis and the CBC substitution approach can help designers to come up with a better structural design that exhibits the satisfactory behaviour in real running state. Although the case study is for a simple table-base model with planary surfaces only, other complex geometries with curved surfaces can be dealt with similarly in terms of analytical procedures. The major problem of dynamic meshing is handled by the extended CBC substitution.

5.5 Discussions

As mentioned above, the finite element analysis for the temperature distribution of the table-base model can be carried out dynamically and continuously with the support by the CBC substitution approach. When a relative motion takes place during the thermal analysis, the CBC substitution takes the responsibility of adjusting the disagreed nodes and guarantees that the analysis can be carried out correctly. In this section, further considerations and concerns of the dynamic thermal analysis are discussed. They are identified so as to make sure that the method is appropriate and valid for the analysis of machines in running state.

From the results of simulations and experiments shown in Figs. 5.5, 5.6 and 5.7, the following are observed:

1. When the table is in running state, the flow of heat is altered due to the relative position of the table and the base. Therefore, some of the temperature curves at the specific points intersect each other. This could be explained by the heat transfers (heat flow within the body by conduction, and heat escape from the body to the surrounding by convection) that vary with the position of the heat source.
2. The number of intersections of two temperature curves depends on the moving speed of the table. The faster the table moves, the more frequent the intersections occur. The number of the intersections is two times of that of the round-trip movements (see Figs. 5.5 and 5.6 for reference). For example, when the table moves at a speed of 3 round-trips/30 min, some of the temperature curves intersect six times.
3. When the table moves faster, the difference in temperature at those specific points of the base becomes smaller. This is because the heat source provides more frequent thermal energy to the base and the conduction has a better chance to spread the energy out to the specific points.
4. It is proved through the case study that the interpolation method is effective to provide the necessary intermediate analytical information for the thermal analysis and to make the calculation continuous. However, this procedure may introduce an unsteady temperature variation depending on the accuracy of the interpolation. One way to guarantee the smooth interpolation is to adopt a higher order polynomial as the interpolation function of nodal temperature values, or to increase the calculation steps in order to shorten the interval needed to be interpolated. Before this is done, the capacity of the computer being used should be taken into consideration for efficiency purpose.

Since there was no effective method known previously to adjust the disagreed nodes when the relative motions take place, the conventional finite element analysis methods could not carry out the thermal analysis of machines in running state, dynamically. It was also difficult for designers to understand the real thermal behaviours of a machine tool once used under actual operating conditions. The experimental methods were so far the general means for this purpose.

With the use of the CBC substitution approach combined with the interpolation method, the *real* simulation of machine tool behaviour becomes possible. Here, the *real* simulation means to simulate the behaviour of machine tools in computers with the same timescale of the actual machining, such as dynamic stiffness, resonance condition of vibration, temperature distribution thermal deformation, etc. These results are meaningful to the optimal design of the machine tools at the detailed design stage.

Whereas the case study reveals the thermal behaviours of the table-base model with a moving heat source, more complicated thermal characteristics remain untouched so far. Generally, there exists a complex phenomenon across the interface between two movable components, where thermal contact resistance exists. When heat flows across the interface, some portion of the heat is lost. As a result, the temperature distribution does not show the continuity across the interface. This is mainly caused by the tiny gaps on the contacting surface due to the roughness and waveness of the components. The heat transfer across the interface can be divided into two parts: (1) heat conduction through the real contacting areas and (2) heat convection through the tiny gaps filled with some sort of fluid (such as coolant or lubricant). Radiational heat transfer can be ignored because of the low temperature difference between the adjacent contacting surfaces. Heat loss is also caused by coolant and lubricant, especially when a machine is moving. It gives more chances to the coolant to take away the heat. Determination of heat loss is also difficult, because it depends on not only the relative moving speed of the components but also the properties of the fluid used. Unfortunately, no theoretical or reliable empirical formula is available. In contrast to the heat loss, some additional heat is generated due to the friction on the contacting surface between moving components. In some cases, the frictional heat source should be taken into consideration as an additional moving heat source for dynamic thermal analysis, if the relative moving speed of the components is fast and the components are heavy.

How to handle this complicated problem is considered critical to the accuracy of a finite element analysis. Instead of using a specific type of finite elements, a concept called *virtual interface* [23] can be applied, across which a virtual heat conduction happens between the contacting surfaces to simulate the heat conduction in reality with heat loss. Frictional heat between relatively moving components can also be taken into consideration as an additional heat source. With this virtual interface introduced to accumulate the complex behaviour across the boundaries of components, the dynamic meshing algorithms introduced in this book become generic and simple. A good convergence in this case study is achieved manually by changing step intervals between calculations. However, this should be based on a well-thought algorithm in the future.

5.6 Concluding Remarks

This chapter deals mainly with how to utilise the CBC substitution in the FEM and how to interpolate the intermediate results of nodal temperature values between consecutive calculating steps in order to ensure the finite element analysis being carried out continuously. Since the thermal problems influence greatly the precision machining, the thermal analysis is, therefore, taken into consideration as an attempt to apply the CBC substitution to dynamic finite element analysis of machine tools.

In this chapter, formulation for the thermal analysis is first carried out to consolidate the foundation of practical calculations. Second, a model used for the thermal analysis is defined and the conditions for calculation are also decided. The model adopted is to represent the table and the base of machine tools, which are the typical components making relative motions frequently during machining operations. Isoparametric elements are considered in this chapter, and an extension of the elements to those with curved boundary surfaces is also presented. The shape functions of 20-nodes isoparametric elements are, third, formulated in order to approximate the actual or exact distribution of the temperature field within an element. Then, the finite element calculation is implemented under several sets of moving conditions and an interpolation method is introduced to handle the intermediate results of the nodal temperature values. Finally, the utilisation of the CBC substitution approach to the FEM is discussed and evaluated against the corresponding experimental results. As a conclusion, the methods introduced in this chapter are appropriate and valid for dynamic analysis with relative motions.

The contents and results included in this chapter are summarised to form the following remarks:

1. Since the thermal behaviours of machine tools are more influential to machining accuracy than other factors, the thermal analysis is taken into consideration in this book as an attempt to apply the CBC substitution approach to the FEM.
2. After the formulation of the thermal analysis with use of the finite element method, the calculation of the temperature distribution is implemented under several sets of moving conditions. The table and base of machine tools are extracted and simplified as the analytic model.
3. From the computational results, it is known that the temperature distributions of the model vary with the position of the heat source and vary with the moving speed of the heat source, too. When the table is in running state, the temperature rises at some specific points are significantly changed depending on the speed of heat source movement as compared to the case with a stationary heat source.
4. Based on the computational and experimental results as well as the discussions, it could be said with confidence that the CBC substitution approach and the interpolation method proposed are extremely effective and appropriate for dynamic analysis like machine tools in running state. The CBC substitution approach is also practically proved to be valid when it is applied to the finite element method for dynamic analysis.

Although the CBC substitution approach is only utilised to the thermal analysis in this chapter, it is expected to be suitable for other types of analyses, such as the vibration, structural and mechanical analyses. Interested readers are encouraged to take one step further by applying the integrated meshing and analysis algorithms towards solving other dynamic engineering problems.

References

1. J.B. Bryan, International status of thermal error research. Annals. of the CIRP **16**, 203–207 (1968)
2. L. Wang, T. Moriwaki, A novel meshing algorithm for dynamic finite element analysis. Precis. Eng. **27**(3), 245–257 (2003)
3. D.M. Hawken, P. Townsend, M.F. Webster, The use of dynamic data structures in finite element applications. Int. J. Numer. Meth. Eng. **33**, 1795–1811 (1992)
4. G. Bedrosian, Shape functions and integration formulas for three-dimensional finite element analysis. Int. J. Numer. Meth. Eng. **35**, 95–108 (1992)

5. T. Moriwaki, N. Sugimura, L. Wang, A modeling system for finite element analysis of machine products. Trans. North Am. Manuf. Res. Inst. SME **21**, 383–390 (1993)

6. L. Wang, T. Moriwaki, An approach to dynamic mesh adjustment for finite element analysis of machines. Trans North Am. Manuf. Res. Inst. SME **24**, 163–168 (1996)

7. L. Wang, Study on methodology of dynamic FEM analysis for integrated CAD/CAE system. in *Proceedings of the 6th IASTED International Conference on Robotics and Manufacturing*, pp. 177–180, 1998

8. M.J. Turner, R.W. Clough, H.C. Martin, J.L. Topp, Stiffness and deflection analysis of complex structures. J. Aeronant. Sci. **23**(9), 805–824 (1956)

9. W. Visser, A finite element method for the determination of nonstationary temperature distribution and thermal deformation, in *Proceedings of 1st Conference Matrix Method in Structure Mech., AFFDL-TR-66-80*, pp. 925–943, 1966

10. E.L. Wilson, R.E. Nickell, Application of the finite element method to head conduction analysis. Nucl. Eng. Des. **4**, 276–286 (1966)

11. O.C. Zienkiewicz, *The Finite Element Method in Engineering Science* (McGraw Hill, London, 1977)

12. C.S. Desai, J.F. Abel, *Introduction to the Finite Element Methods* (Van Nostrand Reinhold Company, New York, 1972)

13. B. Nath, *Fundamentals of Finite Elements for Engineers* (The Athlone Press of University of London, London, 1974)

14. R.T. Fenner, *Finite Element Methods for Engineers* (McMillan, London, 1975)

15. L.J. Segerlind, *Applied Finite Element Analysis* (Wiley, New York, 1976)

16. O. Axelsson, V.A. Barker, *Finite Element Solution of Boundary Value Problems* (Academic Press, Inc., Florida, U.S.A., 1984)

17. H.S. Carslaw, J.C. Jaeger, *Conduction of Heat in Solids*, 2nd edn. (Clarendon Press, Oxford, 1959)

18. G. Yagawa, Introduction of finite element method for flow and heat conduction, Baifukan, pp. 117–118, 1980 (in Japanese)

19. G. Dhatt, G. Touzot, *The Finite Element Method Displayed* (Wiley, USA, 1984)

20. G. Yagawa, S. Yoshimura, Finite element method, Baifukan, pp. 62–69, 1991 (in Japanese)

21. R.J. Collins, Bandwidth reduction by automatic renumbering. Int. J. Numer. Methods Eng. **6**, 345–356 (1973)

22. L. Wang, An interpolating algorithm for dynamic thermal analysis of machine tools, in *Proceedings of the 1st International Symposium on Advances in Integrated, Computer Integrated Manufacturing Systems*, pp. 213–218, 1994

23. T. Moriwaki, L. Wang, Study on thermal behaviour of machine tool elements across movable joints under operating states, in textitProceedings of International Conference on, Precision Engineering, pp. 150–154, 1995

Chapter 6
Conclusions

The major objective of this book is to introduce an integrated system approach for machine tool design, which enables the design/analysis processes of machine tools well-informed, effective and to specifications. Based on this objective, a 3D solid modelling system and a dynamic finite element analysis system are documented. An approach to dynamic finite element mesh generation, called coded box cell (CBC) substitution in this book, is also introduced in order to integrate the engineering analysis with the solid modelling and to share necessary information between the design and the analysis processes, seamlessly with few human intervention. It has been identified through the case studies that the methods presented and the systems developed are appropriate and valid for machine tool design and dynamic analysis under real operating conditions. The contents and results included in this book can be summarised as follows.

In Chap. 2, both the solid modelling methods for three-dimensional objects and the data representing method for machine tool models were discussed. Since the machine tool models can generally be represented as the combinations of a limited number of principal primitives, the constructive solid geometry (CSG) scheme is adopted in this book. Therefore, a low-level data structure is proposed, and seven types of principal primitives are defined for the purpose of establishing a primitive library. Taking the internal structures of machine components into consideration, a new type of primitives—box-type primitives—was introduced to the system, which makes the solid modelling of machine tools and their parts easier and more realistic. Hardware configuration of the system is however left out.

Chapter 3 presents a method to establish a product modelling system for machine tool design. Based on the design process and the information processing in design and analysis of machine tools, considerable requirements to be met in the product modelling system are identified, as well as the definition of a product model. For the purpose of developing such a system, a high-level data structure for both the parts and the entire machine is proposed, which is responsible to represent the geometric and the technological information about the parts of machine tools. The geometric information is described as a set of principal primitives by applying the CSG

L. Wang, *Dynamic Thermal Analysis of Machines in Running State*,
DOI: 10.1007/978-1-4471-5273-6_6, © Springer-Verlag London 2014

modelling technique. The technological information is added to the geometric model as attributes. The information about the physical interfaces and the machine tool structures is represented by graphs, and the information needed for the kinematic simulations is given in the graphs. A product modelling system is presented with use of these techniques, aiming at realising an integrated CAD/CAE system for machine tool design. An interactive kinematic simulation system is also introduced to simulate the kinematic motions of moving parts based on the product models of machine tools. The results of a case study involving kinematic simulation of a vertical machining centre show that the product modelling system is appropriate for representing necessary information during the machine tool design. Finally, relationship between the product modelling system and the integrated CAD/CAE system is outlined.

Chapter 4 proposed a new method for dynamic finite element mesh generation, called the CBC substitution approach in this book. The conventional mesh generation methods are classified, followed by the concept of dynamic mesh generation as the basis for CBC substitution. Hexahedrons are chosen as the mesh elements for the convenience of automatic generation and modification of the finite element meshes, since the machine tools are mostly composed of cuboids and box-type primitives. The basic procedure of dynamic finite element mesh generation can be summarised as follows: (1) generation of the initial hexahedral meshes for the individual primitives and (2) adjustment of element nodes on the interfaces between the adjacent primitives, so as to obtain a FEM model at the same time when its solid model is completed. Full details of the CBC substitution approach is described in this chapter. A new data structure of machine tool model for its utilisation in FEM mesh generation is also proposed, aiming at minimising the computation time for the substitution process. The CBC substitution approach is applied to change the FEM meshes of the primitives at the contacting planes, especially when the relative motions take place. The method presented is validated as effective to generate FEM models of machine tools automatically and dynamically under operating conditions where machine components move relatively. The results of the case studies also demonstrated that the concept of the CBC substitution is simple, logical and straightforward.

Chapter 5 reports some basic applications of the CBC substitution approach to finite element method. This chapter deals mainly with how to utilise the CBC substitution approach in the finite element analysis and how to interpolate the intermediate results of nodal temperature values between calculating steps, in order to ensure the finite element analysis being carried out continuously and correctly. Since the thermal problems influence the machining accuracy significantly, the thermal analysis is, therefore taken into consideration in this chapter to demonstrate how to apply the CBC substitution approach to the finite element analysis. Formulation of the thermal analysis is first performed to consolidate the foundation of practical calculations. A model used for the thermal analysis is defined secondly and the conditions for calculations are also determined. Isoparametric elements are adopted in this chapter, and an extension of the elements to those with curved boundary surfaces is included. The shape functions of 20-nodes isoparametric elements are, thirdly, formulated in order to approximate the actual or exact distribution of the temperature field within an element. Then, an interpolation method is proposed to deal with the intermediate results of nodal temperature values. Based on the results of calculations, discussions and evaluations about the utilisation of the CBC substitution approach to the finite element method are mentioned. Comparisons to real experiments are given to show the effectiveness of the methods. As a conclusion, it is clarified that the CBC substitution and the interpolation methods are effective and appropriate for dynamic analysis, e.g. machine tools in running state. The CBC substitution approach is valid and useful especially when it is applied to the finite element method for dynamic analysis.

The methodologies and results obtained through different case studies are highlighted in this book. Although the CBC substitution is only applied to the thermal analysis of a simplified table-base model, it is applicable to other types of analyses, such as analyses of vibration, construction and mechanics. Interested readers are encouraged to explore this potential to solve real-world problems.

The challenges and problems that still remain to be addressed may be summarised as follows:

1. In the solid modelling system, B-splines technique should be adopted in order to allow the user to specify freeform surfaces of complex products. If a system uses different primitives to approximate these shapes, a large amount of codes is needed to handle combining these primitives into the final object. B-splines provide a solution to this problem.
2. The CBC substitution approach introduced here can be extended by using mapping and inverse mapping techniques in order to be applied to the finite element mesh generation of objects with curved surfaces. However, it is still a challenge for freeform surfaces.
3. The utilisation of the CBC substitution to other types of dynamic finite element analyses should be explored. The interpolation method of the intermediate results of nodal values may be key to obtain satisfactory results. In case that the calculation results are not satisfactary enough, the interpolation method should be adjusted or modified.
4. A database which contains necessary technological information, such as materials, specific gravity, conductivity, etc., should be established to provide detail information for design and analysis.

An integrated CAD/CAE system is expected to be used as a high-performance tool not only for machine tool design but also for design of general-purpose machines. Understanding the true behaviours of those machines in actual running state is the key to reach the high quality of the machines.

About the Author

Lihui Wang is a Professor and Chair of Sustainable Manufacturing in the Department of Production Engineering, Royal Institute of Technology (KTH), Sweden. He received his Ph.D. and MS degrees in Mechanical Engineering from Kobe University (Japan) in 1993 and 1990, respectively, and BS degree in Machine Design (China) in 1982. He was an Assistant Professor at Kobe University and Toyohashi University of Technology (Japan) prior to joining the National Research Council of Canada (NRC) in 1998, where he was a Senior Research Scientist before moving to Sweden in 2008 to take a professorship at University of Skövde. He joined KTH in November 2012.

Professor Wang's research interests are presently focused on energy modelling, dynamic behaviours of machine tools, distributed process planning, web-based real-time monitoring and control, condition-based maintenance, human-robot collaboration, cloud manufacturing and adaptive and sustainable manufacturing systems. His recent work has won a Best Paper Award at FAIM 2002 (International Conference on Flexible Automation and Intelligent Manufacturing) in Germany, a Best Poster Award at IFIP 2003 (IFIP Conference on Virtual Enterprises) in Switzerland, and an Outstanding Paper Award Finalist at NAMRC 2008 (North American Manufacturing Research Conference) in Mexico. He is also an eight-time winner of the NRC Institute Awards on Excellence and Leadership in R&D, Multidisciplinary Collaborative Research, Global Reach and Outstanding People.

Professor Wang has published seven books, three conference proceedings and eight journal special issues. He has authored or co-authored in excess of 240 scientific articles in archival journals, peer-reviewed conference proceedings and books

L. Wang, *Dynamic Thermal Analysis of Machines in Running State*, 127
DOI: 10.1007/978-1-4471-5273-6, © Springer-Verlag London 2014

in the above research areas. In addition to research work, he is actively engaged in various committee and community activities. He was the Conference Chair of FAIM 2004, a member of Grant Selection Committee (GSC-20 for Industrial Engineering) of Natural Sciences and Engineering Research Council of Canada (2004–2007) and a member of Scientific Committee of North American Manufacturing Research Institution (NAMRI) of Society of Manufacturing Engineers (SME) since 2004. Currently, he is the Editor-in-Chief of International Journal of Manufacturing Research, Editor of Robotics and Computer-Integrated Manufacturing, Editor (Northern Europe) of Journal of Intelligent Manufacturing, Associate Editor of Journal of Manufacturing Systems and an Editorial Board Member of other six international journals. He is also a Board Director of NAMRI/SME, a Fellow of SME, an Associate Member of International Academy for Production Engineering (CIRP), a Presidium Member of Swedish Production Academy and a registered Professional Engineer (P.Eng) in Canada.

Printed in the United States
By Bookmasters